# AEC Drafting Fundamentals

Delmar Publishers

*an International Thomson Publishing company* I(T)P®

Albany • Bonn • Boston • Cincinnati • Detroit • London • Madrid
Melbourne • Mexico City • New York • Pacific Grove • Paris • San Francisco
Singapore • Tokyo • Toronto • Washington

## NOTICE TO THE READER

COPYRIGHT © 1994
By Delmar Publishers
an International Thomson Publishing Company

The ITP logo is a trademark under license.

Printed in the United States of America

For more information, contact:

Delmar Publishers
3 Circle, Box 15015
Albany, New York 12212-5015

International Thomson Publishing Europe
Berkshire House
168-173 High Holborn
London, WC1V 7AA
England

Thomas Nelson Australia
102 Dodds Street
South Melbourne, 3205
Victoria, Australia

Nelson Canada
1120 Birchmount Road
Scarborough, Ontario
Canada, M1K 5G4

International Thomson Editores
Campos Eliseos 385, Piso 7
Col Polanco
11560 Mexico D F Mexico

International Thomson Publishing GmbH
Konigswinterer Strasse 418
53227 Bonn
Germany

International Thomson Publishing Asia
221 Henderson Road
#05-10 Henderson Building
Singapore 0315

International Thomson Publishing--Japan
Hirakawacho Kyowa Building, 3F
2-2-1 Hirakawacho
Chiyoda-ku, Tokyo 102
Japan

4  5  6  7  8  9  10  XXX  02  01  00  99  98

Library of Congress Cataloging-in-Publication Data

Chiavaroli, Jules.
    AEC drafting fundamentals / Jules Chiavaroli.
      p.   cm.
    Includes index.
    ISBN 0-314-93452-9
    1. Mechanical drawing.     I. Title.
T353.C5486  1996
604.2--dc20                                           93-48079
                                                        CIP

Dedicated to the memory of John Hamilton, a good friend. He was a casualty of the Vietnam conflict, but he is not forgotten.

Should the students who use this text develop the personal qualities that John had, their contribution to the AEC industry will have a lasting affect.

# Contents

# Contents

# Acknowledgements

I would like to thank my wife, Kathy, and my daughters, Kristin and Lauren, for their patience with me as I worked on this book. They should see more of me now that the work is complete.

Secondly, I would like to thank my colleagues in the Construction Technologies Department, National Technical Institute for the Deaf at Rochester Institute of Technology, who were always happy to answer my questions and give me support. They include:

Hugh Anderson
Jim Jensen
Bob Keiffer
Bill LaVigne
Ed McGee
Yolanda Morley
Ernie Paskey

Finally, I would like to thank the reviewers who took the time to provide me with very valuable feedback along the way.

| | |
|---|---|
| Joe Daudelin | Santa Fe CC (FL) |
| Philip M. Davies | Pasadena City College (CA) |
| Marilyn L. Dietrich | Mercer County College (NJ) |
| Jan C. Fillinger | College of Marin (CA) |
| Thomas R. Freiwald | Delta College (MI) |
| Philip Grau | Milwaukee Area Technical College (WI) |
| Kaffee Kang | Wentworth Inst. of Tech. (MA) |
| James W. Lesslie | Central Piedmont Comm. College (NC) |
| Brian Matthews | Wake Technical College (NC) |
| Fitzhugh L. Miller | Pensacola JC (FL) |
| William J. Morrow | Hillsborough CC (FL) |
| Michael Murphy | San Diego Mesa College (CA) |
| Marguerite Newton | Niagara Co. CC (NY) |
| Phillip Rouble | Sheridan College/Brampton Campus (ON) |
| Peter S. Sabin | West Valley College (CA) |
| Joe M. Samson | Ferris State University (MI) |
| Mike Schnurr | Hillsborough CC (FL) |
| John Silva | Triton College (IL) |
| Thomas L. Turman | Laney College (CA) |
| Harry Vesely | Mesa Community College (AZ) |
| Newton Le Vine | Ramapo College of NJ |
| Daniel G. Winklosky | Guilford Technical College (NC) |

This text was written to assist beginning drafters in making the countless decisions necessary to produce high quality AEC drawings.

The text is broken down into three parts:

• Part I, Fundamentals, introduces students to the basic materials and methods of AEC drafting. Upon completion of this section, students should be comfortable with their equipment and have the basic manual skills necessary to draft well.

Chapter 1 may be broken down into its sections and covered intermittently as students progress through Part I. It may even be beneficial to re-visit this chapter throughout the text.

• Part II introduces students to higher level cognitive skills. It begins with drawing conventions that form the basis of AEC drafting. It concludes with chapters on Pre-Design and Design Drawings, which utilize these conventions.

Upon completion of this part of the text, students should possess the physical and cognitive skills necessary to begin AEC construction drawings in earnest.

• Part III provides a holistic approach to construction drafting by covering the three main disciplines involved in AEC projects. It also includes an overview on the specifics of documentation.

Only by understanding the site, structural, and architectural elements of building projects will students be able to fully comprehend the interrelationship between drawings.

Upon completion of the text and the course associated with it, students should be prepared to tackle larger projects using most of the same skills.

# PART I

# FUNDAMENTALS

# THE AEC INDUSTRY

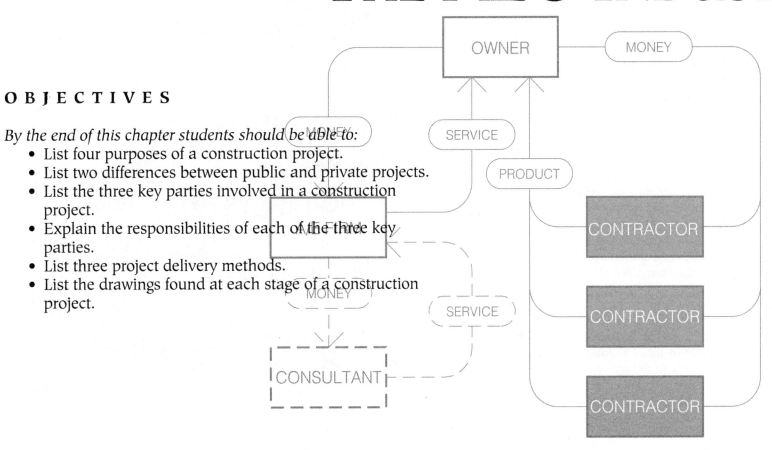

**OBJECTIVES**

*By the end of this chapter students should be able to:*

- List four purposes of a construction project.
- List two differences between public and private projects.
- List the three key parties involved in a construction project.
- Explain the responsibilities of each of the three key parties.
- List three project delivery methods.
- List the drawings found at each stage of a construction project.

1

# THE NEED FOR CONSTRUCTION

The architecture, engineering, and construction (**AEC**) industry is one of the largest industries in the developed world. It produces most of the built environment that surrounds us. Yet we tend to take for granted the industry's products, which include roads, bridges, buildings, waterways, utility lines, tunnels, and even school playgrounds.

The AEC industry satisfies many of our most basic needs and wants. Consider that people cannot live without water, must be sheltered from the elements, must have their waste removed, often travel from one place to another, and generally need power available to them to do their work. All of these needs are at least partially satisfied by the construction industry.

The AEC industry goes far beyond satisfying basic needs however. Its products also enhance and embellish life. For example, a resort complex or a soaring cathedral does much more than provide shelter. A superhighway offers more than basic transportation, and a golf course yields more than simple recreation.

Thus the purpose, or **function**, of a construction project is to satisfy human needs at a variety of levels. A common way to categorize these levels is in terms of physical, intellectual, emotional, and spiritual needs. And indeed, construction projects can satisfy all four.

Each construction project may satisfy a unique combination of functions, and each function may satisfy a range of needs. In this way each project develops its own characteristics.

Everyone involved in a project, including the drafter, should be aware of why it is being developed. Only then can one understand the characteristics, or more appropriately, the **character** of the project. In essence, it is the drafter's job to define this character in a technical way. However, because projects can take years from inception to completion it is easy to "lose sight of the forest among the trees."

**A drafter's first step in a project is to know why it is being built, and to keep this purpose in mind throughout the development of the drawings.** Only then can the drafter create drawings that, when realized, will fulfill the project's purpose.

*Construction projects must satisfy a variety of human needs, from the mundane to the ethereal. A church, for example, satisfies our physical needs by sheltering us from the elements. It satisfies our emotional needs by providing us a place to feel safe. Its interesting design stimulates our intellect, and its majesty uplifts our spirit.*

## PUBLIC VS. PRIVATE PROJECTS

If the first step for a drafter is to know **why** the project is being built, then the next step is to know **who** it is being developed for. One way of categorizing projects is to divide them into public or private projects. **Public projects** are those that benefit the entire population, and **private projects** are those that tend to benefit only the group or individuals who commission them.

Most often, public projects are undertaken to provide a service to the community. Highways, parks, water towers, and government buildings such as libraries and civic centers are examples of public projects.

Private projects such as stores, office buildings, oil refineries, pipelines, and manufacturing facilities are primarily constructed to produce a profit for a business or an individual. If it is not profit, then it is usually some type of gratification that the owners are looking for. This is often the case with a private residence or a religious building.

The distinction between public and private projects is an important one because the procedures and financing of each are quite different. Because these differences affect the preparation of the drawings, they affect the drafter.

Public projects are most frequently developed by government agencies. Libraries, for example, are built by the towns that they serve. Highways are constructed by federal, state, or local highway departments. Projects developed by government agencies on behalf of the public are usually funded with public money, i.e., taxes, bond issues, and the like.

Perhaps the biggest impact on project procedures is that public projects usually must be built using the **competitive bidding** system. This system requires that all parties interested in being hired for the project submit a proposal stating how much they will charge to complete the project as designed. They are basically bidding against each other for the job.

Since drawings are legal documents upon which many companies are staking their future, they must be prepared in the most clear, complete, and professional manner.

Private projects offer more options with respect to how the project can be completed. The owner of a project might be a contractor and would not need as precise a set of drawings as found in the competitive bidding process.

More importantly, private projects are paid for with private money, so the owner can spend the money any way he or she desires. If the money is borrowed, a lending institution will have some influence on the project,

but in general, private projects can be executed in a variety of ways and the drawings are prepared accordingly.

Thus with different motivation and financing found between public and private projects there is a correspondingly different spirit associated with the development and execution of each. The drafter must recognize and understand the differences.

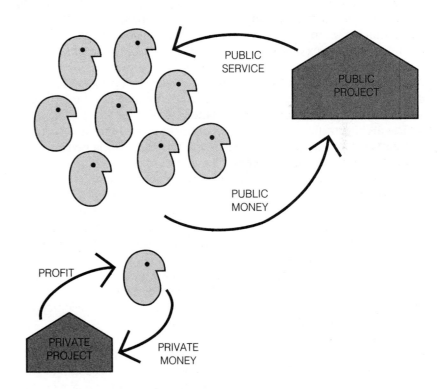

*Private citizens may build projects so as to earn profits for themselves, but when the whole community bands together to create a project it usually benefits the entire community.*

## PROJECT TEAM

To understand the entire building process, it is useful to look at the personnel involved in the AEC industry. In general there are three parties directly related to a project and several others indirectly related. The three main parties involved are:
- owners
- design professionals
- builders

We have already been talking about owners, but we need to define this group a little better. In the construction industry, **owners** are the people who initiate projects. They are the ones who decide that they need a project undertaken, and they may very well be the users of the building.

In many situations the owner is the person or organization that assumes all of the financial risk for a project, but also reaps the benefits if it is successful. For example, the owners of an office park risk financial ruin when they begin such a project, but they stand to make a significant profit if the venture is successful. Often when owners take such financial risks they are also called **developers**. Much of the time developers are not the users of the project but rather the landlords.

The architects and engineers and contractors (the other two parties in a project) tend to call the owner the **client**. The client is the person who hires and pays them and is therefore the focus of their work.

It is possible that developer, client, and owner are all different for a given project. For a school building, the developers would be the superintendent and the school board, the owners would be the residents and tax payers of the school district, and the clients would be not only the school board and superintendent but also the teachers and students because they would need to be consulted about the design and ultimately are the users who need to be satisfied. It is important for the drafter, as part of the design team, to understand the subtle differences between the developer, client, and owner. Perhaps equally important, drafters also need to realize that they are indirectly working for these people.

The design professionals involved in a project include architects, engineers, landscape architects, surveyors, interior designers, technicians, and drafters. Their purpose is to collectively propose an acceptable solution to the owner and then document in detail how the project is to be executed. Throughout the industry and this text, this group is referred to as **A/E**, an abbreviation for architects and engineers.

**Architects** are responsible for designing buildings; **structural** and **civil engineers** design building structures (beams and columns), highways, bridges, and open structures. **Mechanical** and **electrical engineers** design a building's heating/cooling and electrical systems respectively. **Landscape architects** design many of the site features, which include the design and selection of plant materials, exterior furnishings, open structures, and so forth. Civil engineers are also involved with the design of the site. They design drainage systems, paving work, and earth work.

**Surveyors** measure the site before and during construction; **interior designers** select and lay out the furnishings within a building. **Technicians** assist all of the professionals named above by completing drawings and referencing manuals and literature. The term **drafter** has traditionally meant a person who simply produces construction drawings, i.e., they need help obtaining the information being drawn. While this may still be true, employers prefer to see drafters develop the higher level skills possessed by technicians.

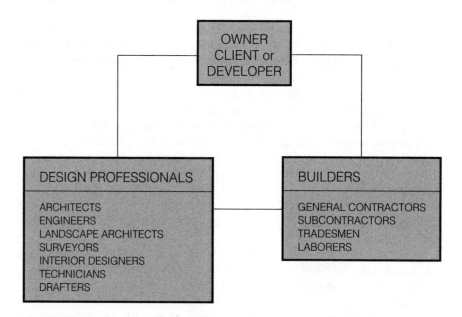

*The three main parties involved in the AEC industry.*

The principal "product" of architects and engineers is a set of drawings, called **working drawings**, which fully explain what needs to be done. There are substantial pages of written information that accompany these drawings, called **specifications**, and together they are the "road map" for getting the project built. Along with legal documents, they are technically called the **contract documents**.

The final party in the construction industry is the group of individuals and organizations that actually construct the project, the builders. There are general contractors, subcontractors, tradespeople, and laborers.

The **general contractor** assumes total responsibility for constructing the project. **Subcontractors** are independent construction firms that specialize in a specific trade such as carpentry, masonry, insulating, etc. They work for the general contractor on a job-by-job basis in much the same way as consultants are hired by A/E firms. **Tradespeople** are the actual skilled laborers who perform the carpentry work, masonry work, etc. They study their trade and work under a mentor, a highly formalized procedure in some trades, not so formal in others. **Laborers** are the assistants to the tradesperson, as the name implies.

Traditionally and in most cases today the three parties introduced here are three separate entities. Increasingly, however, there are alternative arrangements. The exact arrangement has major implications for how the project is executed and how the contract documents are prepared.

Following are the major arrangements found in the industry today for forming a project team.

## Traditional Arrangement

In a traditional arrangement, the owner hires an A/E firm to develop a design to satisfy the needs identified. Part of the service provided by the A/E firm might be a feasibility study or clarifying the owner's needs.

When an overall design is developed to the point of satisfying the owner's criteria, the A/E firm develops the contract documents needed for construction. The A/E firm might also assist the owner in finding qualified contractors to execute the project.

Although the contract documents are a product of the A/E firm, the firm is more basically providing a service that will lead to the ultimate product—the completed project.

A contractor is selected to build the project and is compensated for doing so. The A/E firm continues its service by coordinating with the contractor to ensure that the project is built to the standards defined in the contract documents.

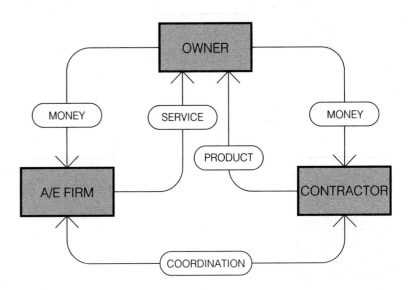

*The traditional arrangement for completing a building construction project includes owner, A/E firm, and contractor. Owners pay the A/E firm to represent their interests to the contractor and then pay the contractor to execute the project.*

## Multiple Design Firms

Sometimes the owner requires the services of more than one design firm. This is the case in many projects because there are often specialized areas of design service that an A/E firm does not ordinarily provide. Thus the firm hires a **consultant** to do that work for them. Acoustical consultants, for example, can help in a theatre design, or a strictly architectural firm could hire an engineer as a consultant to design the structure.

This type of arrangement has ramifications for the drafter. There is a considerable amount of coordination that must occur among consultants. If you are drafting for the A/E firm, you must accommodate the needs of the consultant. If you are a drafter for the consultant, you must rely on information from the main design firm, and you need to work within the framework they establish.

## Multiple Prime Contractors

Another variation from the traditional arrangement is the multiple prime contractor arrangement, also known as the separate contract method.

In the traditional arrangement, the general contractor hires subcontractors to do much of the work. A general contractor's price to the owner includes a percentage to cover the cost of coordinating all of these subcontractors.

If owners want to save this mark-up, they can act as the general contractor themselves. There is money to be saved but enormous responsibility to go with it. Drafters are affected in this method because the information needs to be carefully packaged into distinct, separate pieces rather than in one complete set.

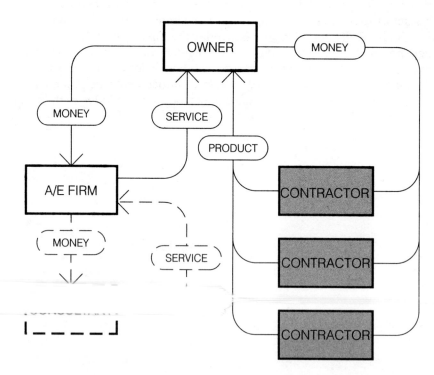

*Some areas of design are very specialized, and a design firm may not offer that capability to the owner. In such cases the primary design firm hires one or more consultants to provide that service to them, and ultimately, to the owner.*

*In this arrangement the A/E firm needs to produce contract documents in multiple packages rather than one complete set.*

## Design/Build

Sometimes an owner finds that dealing with both an A/E firm and a contractor is too bothersome. If money isn't the bottom line, a design/build firm offers a good alternative. Basically one firm will design a project and build it for you, a kind of "one stop shop" for construction projects. This arrangement is also known as **single source**.

A design/build firm has architects, engineers, and contractors in their employ. Dealing with only one firm usually means a higher price but greater convenience to the owner who now has to deal with only one company.

In the preceding arrangements the design work is completed before the contractor is selected. That rather important unknown requires very exacting drawings on the part of the architects and engineers. In contrast, the construction department of a single source firm is brought in at the outset; thus some of the exacting work on the drawings is more relaxed. This is not to say that the quality of the drawings is lower; it's just that there is a built-in, added level of communication between the designers and constructors because they work for the same company, toward the same goal. The design/build A/Es are preparing drawings for their own crews whereas the traditional designers are preparing documents for one or more contractors who have yet to be selected.

## Designer/Developer

A final arrangement of key players is the situation where one entity is owner, A/E, and contractor. This can occur in public or private projects. Some government agencies have their own A/E offices as well as construction crews. With small projects it is fairly common for an agency to do all of the work themselves.

In the private sector, many of the largest real estate projects are now being completed by large companies that design, build, own, maintain, and even finance the entire operation. The daily production of drawings will be highly affected by the individual policies of such organizations.

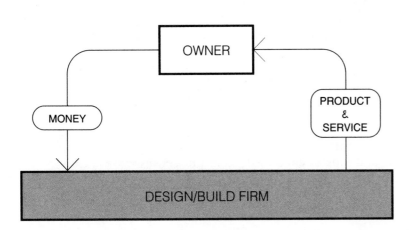

*The easiest arrangement for the owner is this arrangement where total responsibility for the project rests with one firm. It usually comes at a higher price, however.*

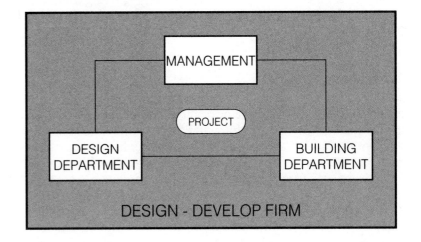

*An organization may have within its means all the resources necessary to design, build, and own a project for profit or for the public good.*

# PROJECT DELIVERY

The arrangement between owner, A/E firm, and contractor is further affected by the method of project delivery. **Project delivery** refers to the manner in which a project is actually constructed. The design and construction phases may be arranged in different ways, each of which has a direct bearing on the way drawings are produced.

## Design-Award-Build

The traditional method of delivering a project is called **design-award-build**. As the name indicates there are three distinct stages in the delivery. It begins with the A/E firm designing the project and producing all of the necessary contract documents. When their work is done, the project goes out to bid and is awarded to the lowest qualified bidder. Finally this bidder constructs the project.

The advantage to this process is that the project goes out to bid. A **bid** is a proposal by a contractor to build the project at a specific price by a specific date. To make it truly competitive, the contract documents need to be very precise. The objective is to have all of the contractors calculate their bids on exactly the same materials and standards of quality.

This delivery method can produce the lowest price because contractors are competing with each other for the project. Only one can get the job so whoever is willing to trim their profit the most should be the lowest bidder.

## Negotiated Contract

In lieu of competitive bidding the owner could negotiate with one contractor from the start. Obviously in this situation the owner loses the advantage of competition, i.e., a lower price. On the other hand, if money is not the most important criterion of the project, selecting a contractor known for quality work can yield a project of very high standards.

This method of project delivery is called a **negotiated contract** because the first cost estimate provided by the selected contractor is frequently too high. The owner and contractor then, with the help of the A/E firm, negotiate back and forth, eliminating or changing material specifications until the price comes within the budget.

## Design-Build

**Design-build** is the method used by the single source arrangement explained in the previous section. Its name is the same as design-award-build except that the word "award" is missing. A contract is awarded, but it is awarded to the single source firm before design begins.

The advantage this method of project delivery has over a negotiated contract is that since the construction department is brought in at the beginning, the negotiation over price and materials occurs along the way rather than at the end of the design stage. This can save time.

The disadvantage of this method is that the architects and engineers act on behalf of the design-build firm. In other methods the A/E firm is the owner's agent and acts on his or her behalf. There is always the question of conflict of interest in the design-build arrangement.

## Fast Track

The **fast-track** delivery method is one that shortens the overall length of project delivery by overlapping the design and construction segments. In other words, construction begins before the drawings are complete.

It can be risky, but it also has its rewards. The project can go up faster, and money can be saved; however, if not managed properly, the process can become a nightmare of errors and cost overruns.

After the overall design has been approved, construction drawings are completed in the same order as construction occurs. Foundation drawings may be completed first and, while that construction work is occurring, structural drawings are being completed.

This process works well for design-build since one company is responsible for everything. The method can be used in a competitive bidding situation as well; it just takes more planning and coordination.

For this option the drafter no longer prepares all drawings simultaneously but rather completes them one phase at a time. Decision making on the drawings must occur more quickly, so accuracy could suffer if not managed properly. There is less time for checking work once construction begins, and subsequent drawings must be sensitive to the changes in the field.

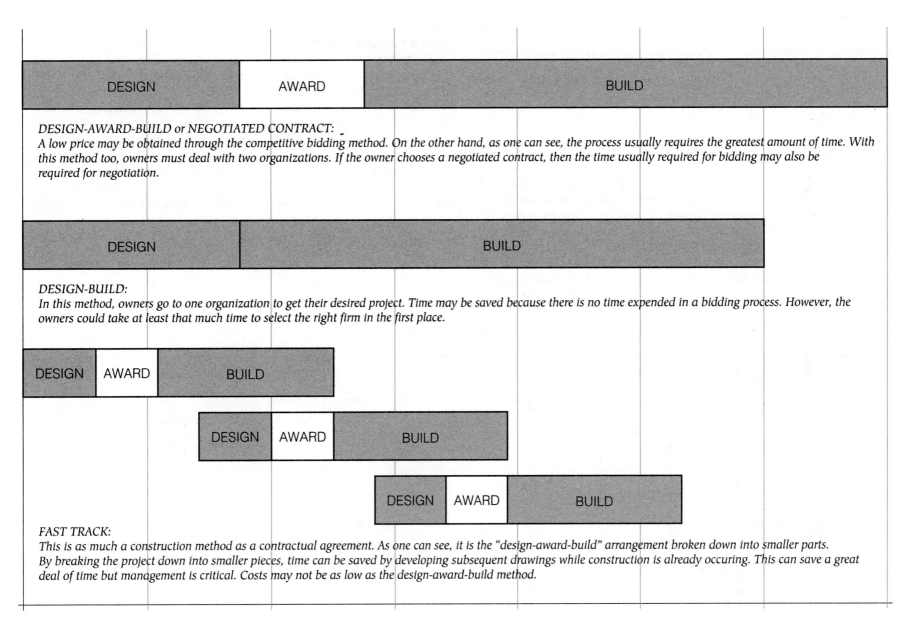

**DESIGN** | **AWARD** | **BUILD**

*DESIGN-AWARD-BUILD or NEGOTIATED CONTRACT:*
*A low price may be obtained through the competitive bidding method. On the other hand, as one can see, the process usually requires the greatest amount of time. With this method too, owners must deal with two organizations. If the owner chooses a negotiated contract, then the time usually required for bidding may also be required for negotiation.*

**DESIGN** | **BUILD**

*DESIGN-BUILD:*
*In this method, owners go to one organization to get their desired project. Time may be saved because there is no time expended in a bidding process. However, the owners could take at least that much time to select the right firm in the first place.*

**DESIGN** | **AWARD** | **BUILD**

**DESIGN** | **AWARD** | **BUILD**

**DESIGN** | **AWARD** | **BUILD**

*FAST TRACK:*
*This is as much a construction method as a contractual agreement. As one can see, it is the "design-award-build" arrangement broken down into smaller parts.*
*By breaking the project down into smaller pieces, time can be saved by developing subsequent drawings while construction is already occuring. This can save a great deal of time but management is critical. Costs may not be as low as the design-award-build method.*

TIME

# PROJECT STAGES

All projects, no matter what the method of delivery, go through a five-stage process: pre-design, design, bidding and negotiation, construction, and post-construction. It is important to have a good understanding of each stage as well as the purpose or intent of the drawings associated with each stage.

## Pre-Design

The process begins with an individual or organization discovering a need and initiating the project by hiring an A/E firm or design/build firm to assist them. The first step is for the owner to convey to the designer exactly what is needed. Together, after much discussion, the needs are formally organized into what is called a *program*.

Without a program it would be impossible for the designers to do their work. A program contains all of the particulars of what the owner wants, and in essence, defines the problem(s) that the design will solve.

For example, a program for a home would state the number of spaces needed, whether a formal dining room or eat-in kitchen is desired, and even what feeling the design should evoke such as "light and airy" or "solid and strong."

Usually drawings are not involved in the pre-design stage, but sometimes there is already a structure on the site, e.g., the project is merely an addition or perhaps the project is the renovation of an existing structure. If there are no drawings on record for such structures, then the designers need to survey what is there and produce a set of *measured drawings*.

In a similar fashion, for virtually every project, a *site survey* must be done of existing site conditions. This is the responsibility of a land surveyor, either hired by the A/E firm or already part of their organization. In each of the above cases the key is that the drawing(s) show precisely what is at the site. Commonly this is also called a *base map*.

| PROJECT STAGES | PRE-DESIGN | | DESIGN | | | BID. & NEGOT'N. | CONSTRUCTION | | POST CONSTR. |
|---|---|---|---|---|---|---|---|---|---|
| DRAWING TYPES | site survey | measured drawings | schematic design | design developm't | working drawings | working drawings | shop drawings | working drawings | as-built drawings |
| Surveyor | ● | | | | | | | | |
| Architect | | ○ | ● | ● | ● | ○ | | ○ | ◐ |
| Landscape Architect | | ○ | ● | ● | ● | ○ | | ○ | ○ |
| Civil, Structural, Mechanical and Electrical Engineers, Interior Designers | | ○ | ◐ | ● | ● | ○ | | ○ | ◐ |
| Contractor/Subcontractors | | | | | | | ◐ | | |
| Fabricator/Supplier | | | | | | | ● | | |

● = usually needed    ◐ = sometimes needed    ○ = not usually needed

## Design

The next stage of the process is the design stage. Architects articulate this stage more than the other disciplines mainly because their type of design is extensively creative as well as technical. In explaining this stage, we will use the three phases followed by the architectural profession and generally referred to as the:
- schematic design phase
- design development phase
- contract document phase (working drawings and specifications)

The first step is the brainstorming part of design called the *schematic design phase*. Many ideas are explored in a short amount of time.

One design from all the schematics usually emerges as the strongest and *design development* drawings are produced from it. It is in these drawings that the architect begins to solve the technical aspects of the design and where the consultants come together to coordinate their efforts.

The design stage is culminated with the production of **contract documents**, the two parts being **working drawings** and **specifications**. These are solely technical drawings and written descriptions for the builder to execute the project.

At this point it would be appropriate to discuss interior design drawings. Interior design relates to most of the equipment and furnishings that are not attached to a structure. For many projects this is done by the owner so an interior designer is not used. In commercial projects however, it is common to have a set of interior drawings accompany the working drawings. Specifications are also written. The drawings usually include plans showing locations for all these items.

## Bidding and/or Negotiation

If the project is not utilizing the design-build method, then a bidding and/or negotiation stage is necessary. At this point, depending on cost and budget, the decision is made whether to proceed, and if so, who is awarded the contract. No drawings are produced at this stage except occasionally when the A/E firm needs to issue revised working drawings to clarify some aspect of the design.

## Construction

Once the contract is awarded, actual construction begins. Throughout the construction phase *shop drawings* are produced by the suppliers and fabricators of various materials and by subcontractors in order to further detail work shown on the working drawings. These are submitted to both the contractor and the appropriate design professional for approval.

Contractors do on occasion produce some drawings for their own benefit during construction. They often have personnel who can draft whatever drawings are necessary to help the tradesperson complete the work.

## Post-Construction

Upon completion of construction, the owner takes possession of the project and may receive a set of *as-built drawings* as part of the contract. This set of drawings delineates the way the project was actually built as opposed to how the project was drawn in the working drawings, since changes at the site are inevitable. Most often the A/E firm(s) that completed the working drawings also complete the as-builts.

# Drawing Types

The phases and stages presented in the previous section may be somewhat confusing to a novice; however the drawings involved can be broken down more simply. For the purposes of this text, there are three general categories of drawings that are most important, namely:

- pre-design drawings
- design drawings
- construction drawings

The purpose of pre-design drawings is to record what the existing conditions are. These drawings are needed by the architect and engineers to begin the design process.

The goal of design drawings is to illustrate the design to the client, potential investors, other design team members, and the local planning board. Schematic design drawings are completed first, followed by design development drawings.

Finally, construction drawings are produced as a set of "instructions" for the contractor. They are the technical drawings that show how the project is to be built.

## Pre-Design Drawings

### Site Survey

The *site* is the land where the project is to be built. Before anyone can begin to design a project, it is necessary to ascertain exactly what is there by conducting a site survey. A base map of the property must be drawn from which the design can spring.

Surveyors are licensed professionals who do the survey and record their findings for others to use. Their job involves researching what is already on file about the site at the local town hall and using surveying instruments at the site to finish the job.

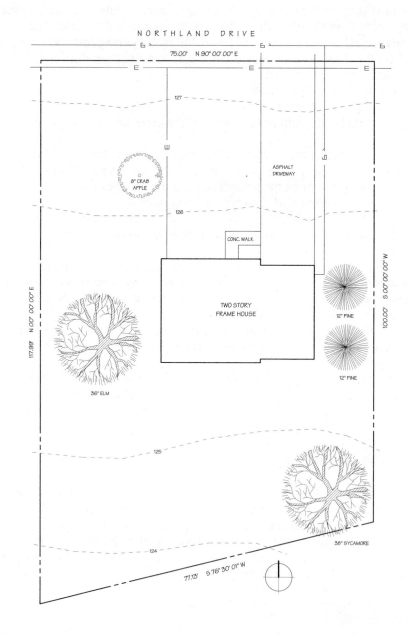

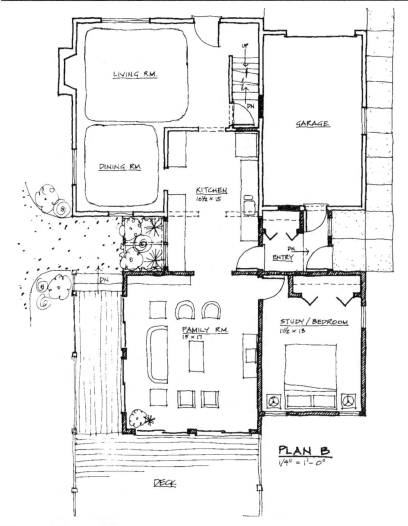

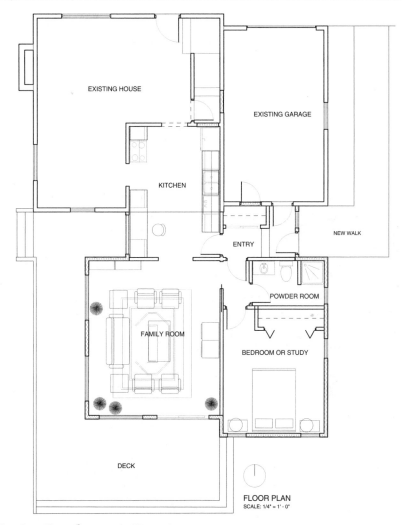

## Design Drawings

### Schematic Design Drawings

During the design phase many possible solutions to the design problem are explored. It is when all the ideas are boiled down to the best one, two, or even three that the architect begins to share them with the client. With continuing dialog, re-design, and hard decision making one design emerges as the best solution for the client's needs.

## Design Development Drawings

Once preliminary approvals are obtained, the project takes on a much more definitive nature. The architects and engineers enter into the design development phase where all the technical aspects of the project are addressed.

The most important function of design development work is the coordination among all the team members. It is through design development drawings that the architects and engineers assure themselves that everything will fit together and work properly.

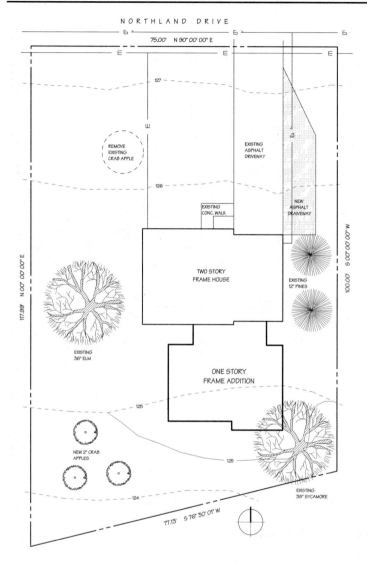

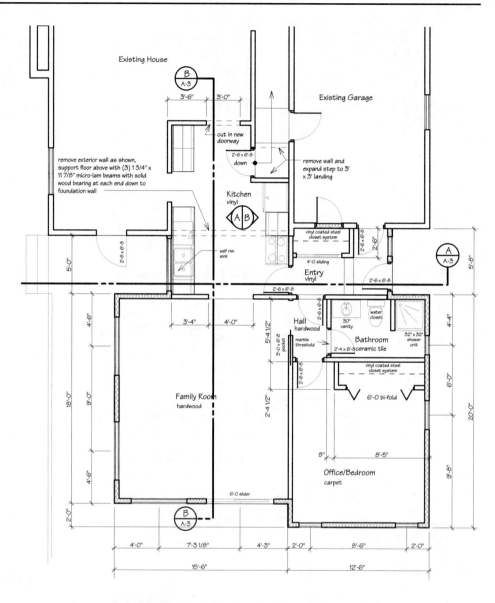

## Construction Drawings

### Civil/Site Working Drawings

The civil/site drawings explain what aspects of the site are to
remain unchanged, which are to be removed, and finally what
additions are to be made to the site.

### Architectural Working Drawings

The architectural construction drawings have evolved to complete technical
documents at this stage. The building's walls, floors, and roof are completely
defined and detailed.

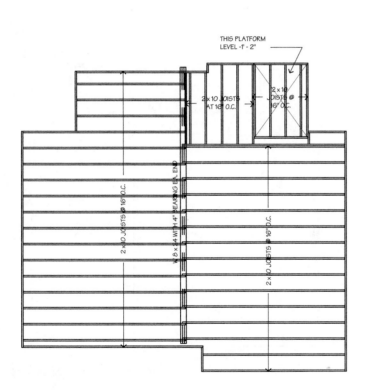

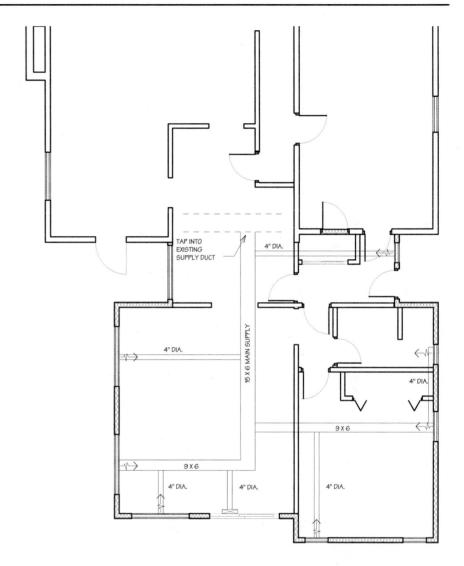

## Structural Working Drawings

Structural construction drawings define the "bones" of the building, the footings, beams, columns, bracing, and decking required to support the building and the other sub-systems.

## Mechanical (HVAC) Working Drawings

Some mechanical engineers specialize in the heating, ventilating, and air conditioning (HVAC) systems needed in a building. Their job is to design and draw the mechanical equipment necessary for such a system, including the size of ducts, where they go, and details of all of the equipment controls.

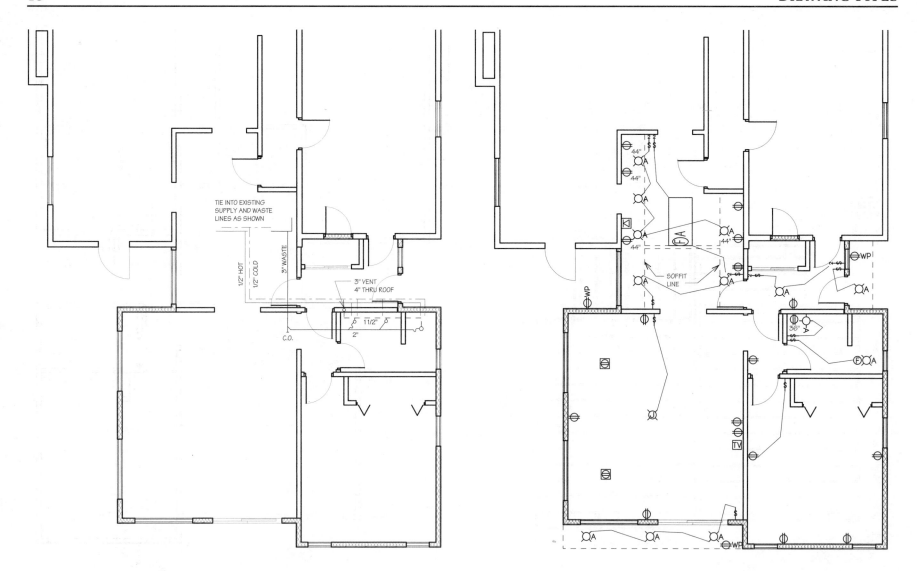

## Mechanical (Plumbing) Working Drawings

Other mechanical engineers specialize in plumbing systems for buildings. Their working drawings detail fixtures, controls, and piping related to the plumbing. Even more specialized are those who design only fire protective plumbing systems.

## Electrical Working Drawings

Electrical engineers must design and draw the power equipment, wiring, controls, and lighting needed to satisfy the program requirements.

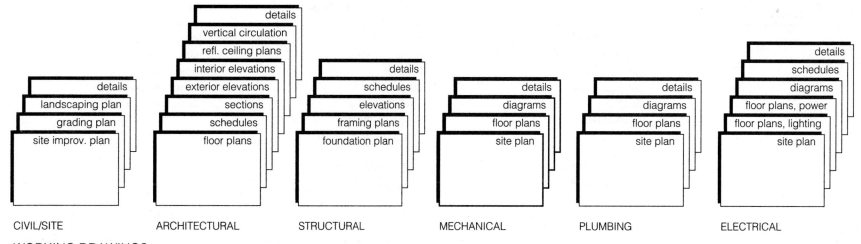

**WORKING DRAWINGS**

Labels from the drawing stacks:

CIVIL/SITE: details, landscaping plan, grading plan, site improv. plan

ARCHITECTURAL: details, vertical circulation, refl. ceiling plans, interior elevations, exterior elevations, sections, schedules, floor plans

STRUCTURAL: details, schedules, elevations, framing plans, foundation plan

MECHANICAL: details, diagrams, floor plans, site plan

PLUMBING: details, diagrams, floor plans, site plan

ELECTRICAL: details, schedules, diagrams, floor plans, power, floor plans, lighting, site plan

## CONTRACT DOCUMENT SET

Working drawings constitute the greatest single drafting effort during the cycle of the project. These drawings are produced at the end of the design stage by all of the architects and engineers involved and are arranged into sets as illustrated above.

It is important to remember that working drawings are only part of the contract documents. The other part is the written portion called the *project manual*. The project manual is a text that includes the following:
- **bidding requirements,** which are instructions to help contractors submit proper bids
- **contract forms, bonds, and certificates,** which are quality assurance documents
- **conditions of the contract**, which are general legal documents
- **specifications**, which are the actual material descriptions and installation procedures

The first three items are essentially legal papers that tie together the contract documents and set the ground rules that the bidders and eventual contractor must follow. They identify who is responsible for what and how payments are to be made. They also protect each party to some degree.

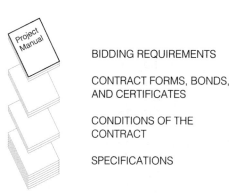

Project Manual

BIDDING REQUIREMENTS

CONTRACT FORMS, BONDS, AND CERTIFICATES

CONDITIONS OF THE CONTRACT

SPECIFICATIONS

PROJECT MANUAL OR "SPECIFICATIONS"

The specifications themselves are the detailed instructions that explain exactly what must occur on the project itself. This part is so much the heart of the project manual that the manual is most commonly referred to as the specifications.

The role of each part of the contract document set is quite clear. The drawings are supposed to show "what" and "where." The drafter needs to show the contractor what each material is and where it is to be placed. The specifications are to explain "how" and define "quality." They tell the contractor the manner in which to install or apply materials and what level of quality the materials are to be.

# KEY TERMS

AEC
function
public project
private project
competitive bidding
owner
developer
client
A/E

architect
structural engineer
civil engineer
mechanical engineer
electrical engineer
landscape architect
surveyor
interior designer
technician
drafter
working drawings
specifications

contract documents
general contractor
subcontractor
tradesperson
laborer
consultant
single source
project delivery
design-award-build
bid
negotiated contract
design-build

fast-track
program
measured drawings
site survey
base map
schematic design
design development
shop drawings
as-built drawings
site
project manual

# PRACTICE EXERCISES

1. Identify the purposes of the following construction projects:

   parking lot
   stadium
   city street
   communication tower
   garage

   amusement park
   shopping mall
   city sidewalk
   bridge
   airport

2. Identify whether the following are public or private projects or both:

   interstate highway
   electrical utilities
   university
   shopping mall
   religious school

   corporate headquarters
   library
   golf course
   home
   motor vehicle office

3. List at least seven of the key parties involved in a construction project and explain the role played by each of them.

4. Explain the chronological sequence of events in a construction project from beginning to end, indicating where each of the people listed in number 3 above is involved in the sequence.

5. Identify which project team arrangement and delivery method(s) is/ are best suited for the following situations and explain your answers:

   the owner wants the fastest delivery
   the owner wants the lowest cost
   the owner wants the most convenient working relationship
   the owner wants the highest quality

6. Identify in which stage of the construction project the following drawings are prepared:

   design development
   shop drawings
   working drawings
   schematic designs

   site survey
   as-built drawings
   measured drawings

7. Identify which drawings are produced during the following project stages and who produces them:

   pre-design
   design

   construction
   post-construction

8. List the correct order in which the following drawings are completed: design development, site survey, shop drawings, as-built drawings, working drawings, measured drawings, schematic designs.

# 2

# DRAFTING EQUIPMENT, SUPPLIES, AND TECHNIQUES

**OBJECTIVES**

*By the end of this chapter students should be able to:*
- Identify basic manual and computer aided drafting equipment and supplies.
- Explain the function of each of the basic drafting equipment and supplies.
- Correctly line up and tape a sheet of drafting media to a drafting board.
- Construct a drawing using various line types in the correct sequence.
- Construct a drawing from top to bottom and from right to left (left to right for left handers).
- Measure lines exactly along the axis to be drawn.
- Draw various lines, which meet or intersect to standards established in this chapter.
- Draw lines at 15 degree increments using the two standard fixed triangles.
- Draw a compound curved line using French curves.
- Use lead and ink according to standards established in this chapter.

# MANUAL DRAFTING EQUIPMENT AND SUPPLIES

In today's drafting room there are more tools available then ever before. And although computers are being used increasingly, they have yet to completely replace manual tools in AEC offices. In architectural offices, in particular, many believe that there will always be the need for manual tools and techniques.

Thus, presented here are the tools and techniques for both manual and computer aided design/drafting.

*Drafting boards are available in a wide variety of styles. Most are already attached to a table as shown here. Features include an adjustable top, drawers, foot rest, pencil catch at the bottom, and durable surface. Most are made of either wood or metal.*

DRAFTING BOARD: The starting point for a drafter's workstation is the **drafting board.** Most are equipped with legs to stand on the floor. Those without legs are designed to sit on a table. In either case the board should be stable and strong since it will be heavily used.

SURFACE COVER: A separate but important part of the board is a vinyl **surface cover**. This cover provides a smooth and slightly resilient base upon which to draft. It can be held in place with double sided tape and is usually light colored.

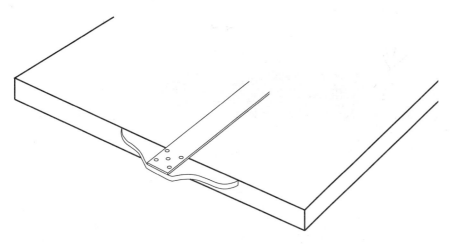

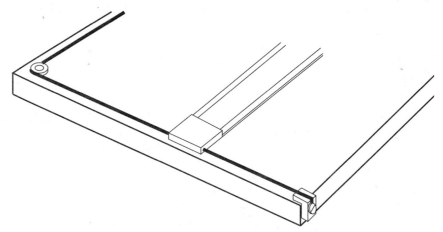

**T-SQUARE:** This is a free sliding instrument that moves up and down on a drafting board. The drafter keeps it parallel by keeping the "T" end flush against one of the edges of the board. It establishes a horizontal base for drawing lines or for orienting other equipment. It has been largely replaced by the drafting machine and parallel rule.

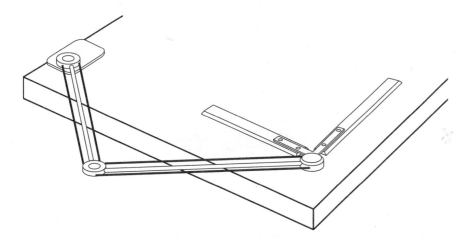

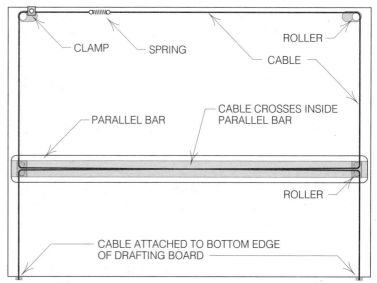

**DRAFTING MACHINE:** This device is more popular in engineers' offices than in architects'. It is attached to the drafting board and has integral vertical and horizontal scaled straight edges. The device moves about the board, providing the opportunity to draw vertical and horizontal lines. The straight edges may be rotated and locked, providing the drafter with the ability to draw any angle.

**PARALLEL RULE:** Also called a parallel bar, this device is attached to the drafting board with thin wire and moves up and down, staying parallel to its original position at all times. It too is used for drawing horizontal lines and as a guide for orienting other drafting equipment.

A drafter should be prepared to replace the cable in a parallel rule at any time since they occasionally snap. The roller covers at either end of the bar can be removed for easy access. The wire runs around the board as shown above, guided by rollers but clamped to the board at the top. The parallel rule moves while the wire acts as a guide.

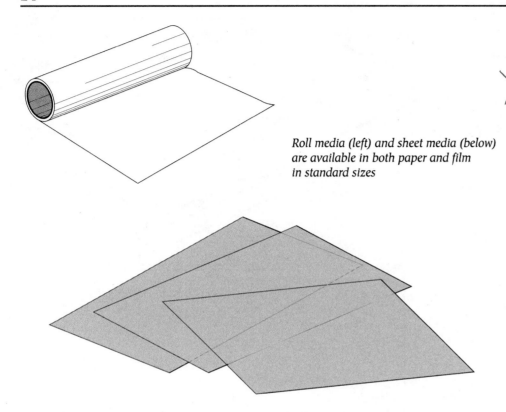

*Roll media (left) and sheet media (below) are available in both paper and film in standard sizes*

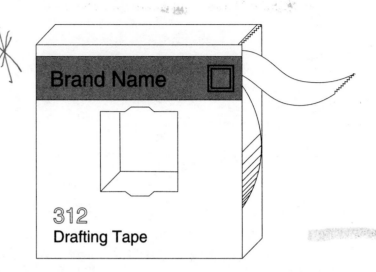

DRAFTING MEDIA: Drafting media is the actual material on which a drafter puts lines. Media is available in rolls or sheets and comes in various qualities and thicknesses. It can generally be divided into two categories, papers and films. Both are translucent to allow light to pass through in order to make reproductions.

Paper is less costly, the cheapest of which is called **bumwad**, **onion skin**, or **tracing paper**. It is usually yellow or white in color and comes in roll form. This lowest quality paper is most suitable for early sketches and preliminary drafting. Higher in quality is another tracing paper called **vellum**. This is white in color and of heavier weight (thickness); 20 lbs. is the standard for general purpose drafting. It is available in both roll and sheet form.

The higher quality medium is **drafting film**, commonly called by the brand name **Mylar**®, which is duPont's variety of the product. The advantages of film are its durability and erasability. This of course comes along with a higher price. The glossy film base itself is not suitable for drafting so it is coated with a satin finish known as the **tooth**. Film can be obtained with tooth on one or both sides.

Both paper and film will accept lead and ink. Each requires its own variety of leads, inks, and erasers as we shall see in the next few pages.

DRAFTING TAPE: Drafting media is held to the board with drafting tape. This is not to be confused with masking tape, which has more adhesive. Masking tape can damage drafting media when it is removed. Drafting tape is designed and manufactured specifically to hold drawings to a drafting board and is obviously preferred.

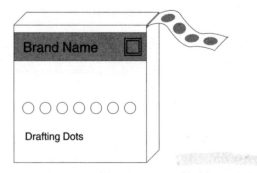

DRAFTING DOTS: Drafting dots and strips are pre-cut pieces of drafting tape mounted to a paper backer and rolled into a box. They will dispense easily and have become very popular in recent years.

**LEAD HOLDERS:** Drafters use traditional wood pencils for drafting but more often use *lead holders* and *mechanical pencils*. Wood pencils are often used for lettering and layout work, are readily available, and are easily sharpened with a standard pencil sharpener. Lead holders are basically hollow metal or plastic tubes that hold leads in place. They are flexible in that different leads can be interchanged, the point of the lead can be honed to any shape, and the point can project at any length from the body of the holder.

Traditional lead holders accommodate leads about the same diameter as is found in a wooden lead pencil. They require ongoing sharpening to maintain consistent line width. Mechanical pencils are the same as lead holders except that they hold much thinner leads, from .3mm to .9mm in diameter. They can maintain the same line width without sharpening. However, this advantage is somewhat offset by the mechanical pencil's thinner lead, which tends to break easily.

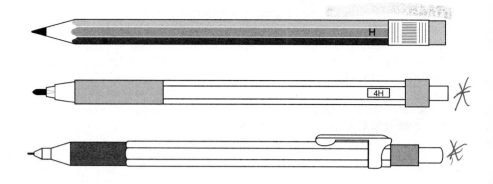

*From top to bottom: the standard wood pencil, the lead holder, and the mechanical pencil. Each has its place in the drafting field, and often each drafter develops personal preferences for their use.*

**DRAFTING LEADS:** The leads that fill lead holders are made of graphite to make lines on paper media. In general a hard lead will produce a thin and light line; a soft lead will produce a thicker and darker one. The range of leads from hardest to softest is 9H, 8H, 7H, 6H, 5H, 4H, 3H, 2H, H, HB, F, B, 2B, 3B, 4B, 5B, 6B.

Plastic-based leads are made for films. They range from hard to soft, designated E5, E4, E3, E2, E1, E0 by one manufacturer, and N5, N4, etc., by another manufacturer. The thin leads required for mechanical pencils are made from a combination of synthetic polymer resin, graphite, and carbon for the purpose of resisting breakage. They are available for both papers and films and carry the same designations as above.

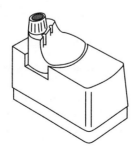
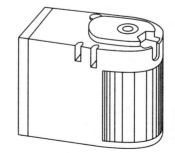
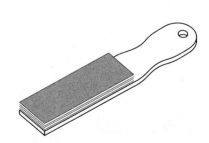

**LEAD POINTERS:** There are lead pointers to fit all budgets and needs. The purpose of the lead pointer is to hone the end of the lead to the perfect shape for producing a specific line thickness. The more expensive lead pointers can vary the sharpness of the point to do this. Some are manually operated while others are electric. Sandpaper pads are useul for shaping leads by hand.

**TECHNICAL PENS:** The drafter uses a technical pen to make lines with ink. The pen is usually made of a hollow plastic tube which holds the ink. Attached to the drawing end is a metal tip, in essence a thin tube, which allows the ink to flow in a very consistent line width. The pen points are removable for cleaning and/or replacement and are available to produce line thicknesses from .13mm to 2mm.

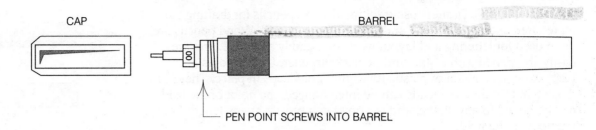

CAP          BARREL

PEN POINT SCREWS INTO BARREL

**INK:** Drafting ink is generally waterproof and dense black, qualities that ensure permanent lines on drawings. Several varieties and colors are available. There is one mixture for papers, another for films, and yet one more specifically for very small pen tips. Ink comes in small plastic or glass bottles.

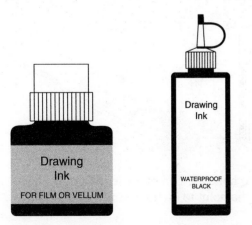

| PEN NO. | METRIC | LINES PRODUCED BY PEN POINTS NAMED AT LEFT |
|---------|--------|--------------------------------------------|
| 6 x 0 | .13 mm | |
| 4 x 0 | .18 mm | |
| 3 x 0 | .25 mm | |
| 00 | .30 mm | |
| 0 | .35 mm | |
| 1 | .50 mm | |
| 2 | .60 mm | |
| 2.5 | .70 mm | |
| 3 | .80 mm | |
| 3.5 | 1.00 mm | |
| 4 | 1.20 mm | |
| 6 | 1.40 mm | |
| 7 | 2.00 mm | |

LINE WIDTHS VARY DEPENDING ON MEDIA, INK, HUMIDITY, SPEED AT WHICH THE LINE IS DRAWN, ETC.

SCALES: There are three types of scales used to measure drawings: the ***architect's scale***, the ***engineer's scale***, and the ***metric scale***. The architect's scale measures in feet and inches at the ratios of 3/32" = 1' - 0" to 3" = 1' - 0". The engineer's scale measures feet and tenths of a foot at user-defined ratios based on 10, 20, 30, 40, 50, and 60. Thus the user could set 1" = 20', 1" = 200', 1" = 2000', etc. Metric scales are based on the metric system, which uses the meter and subdivides into tenths (decimeters) and hundreths (centimeters). Common scales are 1:10, 1:20, 1:30, etc. All are available in wood, metal, and plastic.

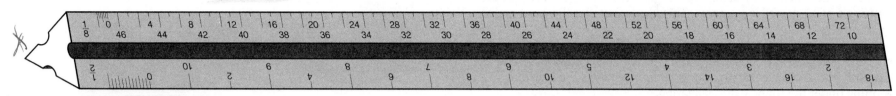

ARCHITECT'S SCALE

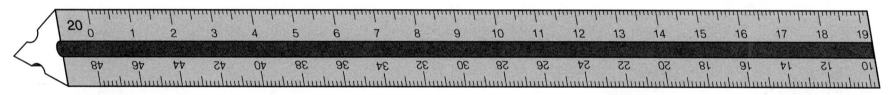

ENGINEER'S SCALE

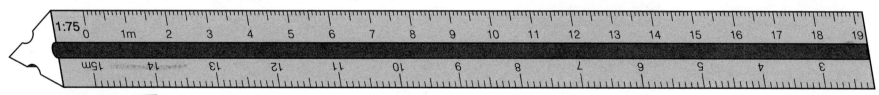

METRIC SCALE

SOLID TRIANGLE

HOLLOW TRIANGLE

2 BEVEL FLAT

OPPOSITE BEVEL

4 BEVEL FLAT

TRIANGLES: Triangles are used with a parallel rule to draw lines at angles. They are available in a variety of sizes but in only three basic configurations. The *adjustable triangle* can be, as its name implies, adjusted to any angle by moving one of its sides and locking it into place.

The other two types of triangles are fixed. One is the *30-60-90 triangle*, and the other is the *45-45-90 triangle*. Their names indicate the angles they define. By using them together, the drafter may produce any angle at 15 degree increments. This is adequate for the great majority of projects.

Triangles are available with two kinds of edges. Those with a square edge are made for use with lead. Those with an offset edge are made for use with ink. The offset for an inking triangle prevents the ink from running under the triangle and smearing.

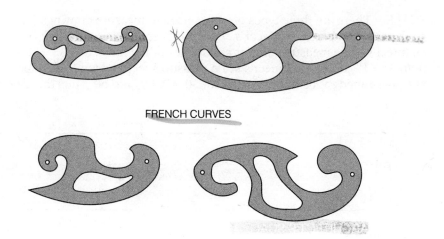

FRENCH CURVES

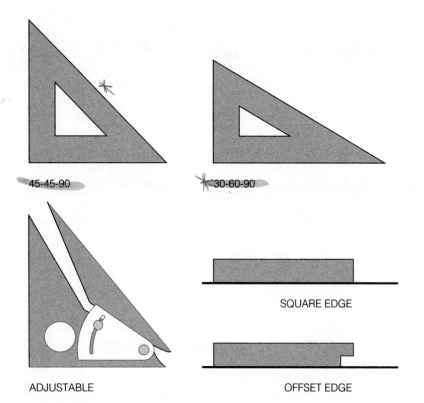

45-45-90

30-60-90

ADJUSTABLE

SQUARE EDGE

OFFSET EDGE

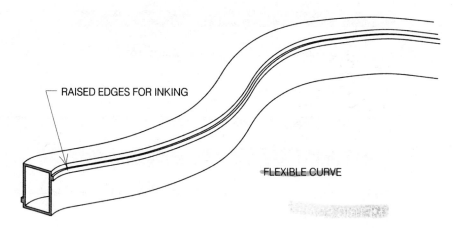

RAISED EDGES FOR INKING

FLEXIBLE CURVE

CURVES: Curves are used to draw irregularly curved lines smoothly. Those that are rigid are irregular curves, or *French curves*. This tool requires the drafter to draw a curved line in several sections. Since it is impossible to make a template for every curve at every scale a drafter might encounter, curves are available in many sizes and shapes.

A *flexible curve* differs from the French curve in that it is thicker, is made of a flexible plastic material, and can be bent into any shape the drafter desires. Once shaped it can be used as a drawing edge just like the rigid curves above.

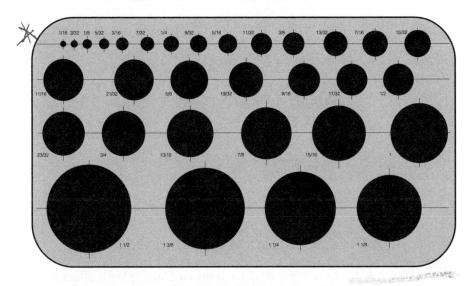

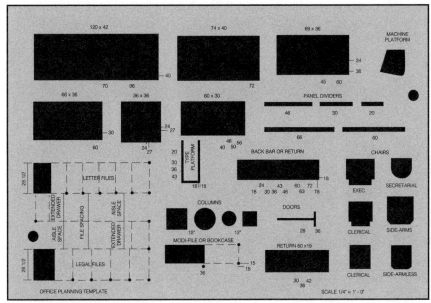

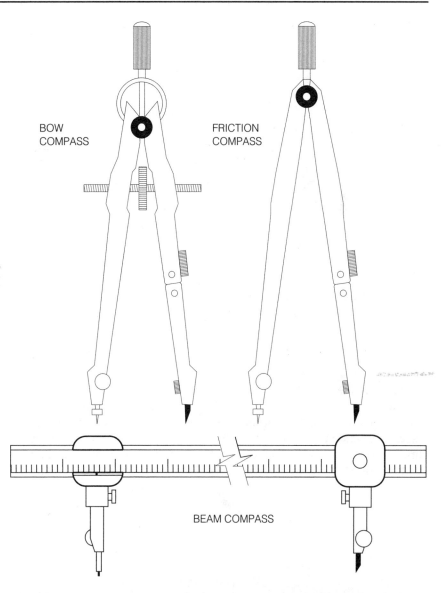

BOW COMPASS

FRICTION COMPASS

BEAM COMPASS

TEMPLATES: Templates are punched plastic drafting tools that may be traced to produce commonly used shapes. The most common one is the circle template, which eliminates the need for a compass. Others include other geometric shapes, symbols, lettering, plumbing and electrical symbols, furniture, and so forth.

COMPASSES: Compasses are drafting instruments that help the drafter draw circles. They may be loaded with lead or pen points. **Bow** and **friction compasses** will produce small to medium-sized circles. Large circles and arcs may be done only with a **beam compass**. This device is a long beam on which a drafter can attach and adjust a pin at one end and a lead or pen point at the other.

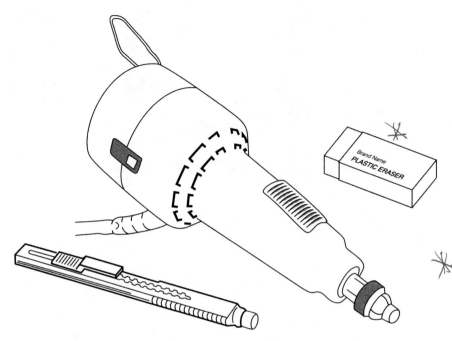

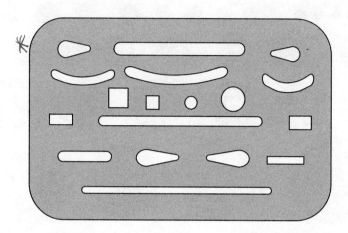

**ERASING SHIELD:** An erasing shield is a thin metal or plastic sheet that resembles a template. Its function is to allow the drafter to erase only that part of the drawing that is exposed and protect everything around it. The thinner the material the better, which makes metal erasing shields superior to plastic ones.

**ERASERS:** Erasers are used to remove unwanted lines from drawings. They are made from various compositions to erase the many types of leads and inks from papers and films. The rubber type erasers (such as Pink Pearl®) will remove graphite lead from paper, whereas plastic erasers are made to remove plastic lead and even ink from vellum and film. Each type of eraser may be found in the familiar hand size cube or in stick form to fit inside manual holders and electric erasing machines.

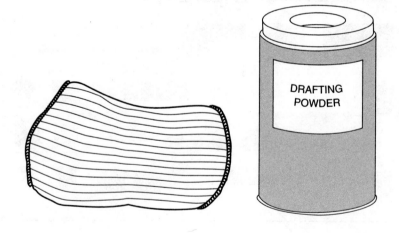

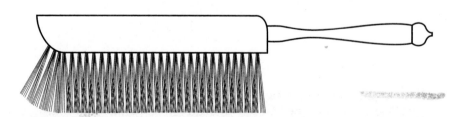

**DRAFTING BRUSH:** A drafting brush is a simple but necessary tool. It is used to clean erasures and graphite from the work surface in order to keep drawings clean.

**DRAFTING POWDER:** This white powder helps prepare the surface of the media to accept ink better. A second type is available to help keep the media clean while working with lead. It is sprinkled on the media and can be spread by moving equipment. This second type most often comes in a porous cloth bag.

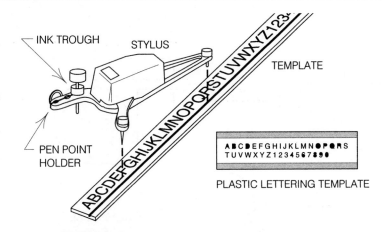

INK TROUGH

STYLUS

TEMPLATE

PEN POINT
HOLDER

PLASTIC LETTERING TEMPLATE

**LETTERING TEMPLATE:** The most common lettering templates are *Leroy*® lettering sets, which include a stylus. They are favored in engineering offices more than in architectural offices. Plastic lettering templates are also available.

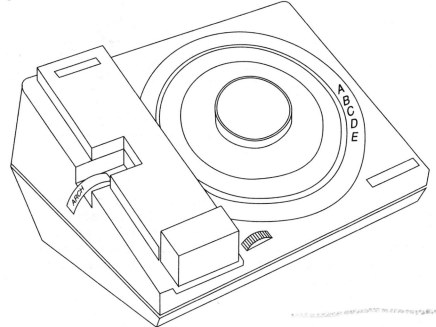

**LETTERING MACHINES:** Machines by Kroy® or Merlin® impact a ribbon to transfer letters onto a tape with adhesive backing. The lettering can then be transferred to a drawing.

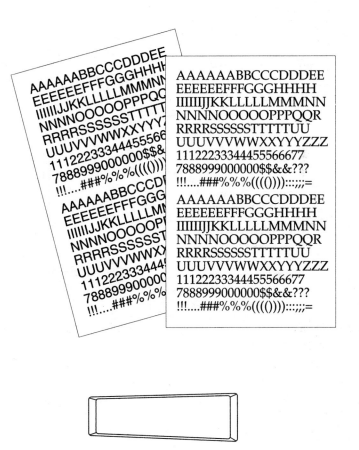

BURNISHING TOOLS

**TRANSFER LETTERING:** These are sheets of preformed letters and symbols. The drafter positions the sheet where desired and then rubs off a letter using a blunt instrument. The letter is transferred to the drafting media where it must then be burnished. *Burnishing* is done by placing a wax paper over the letter and then rubbing hard to make sure the letter adheres properly.

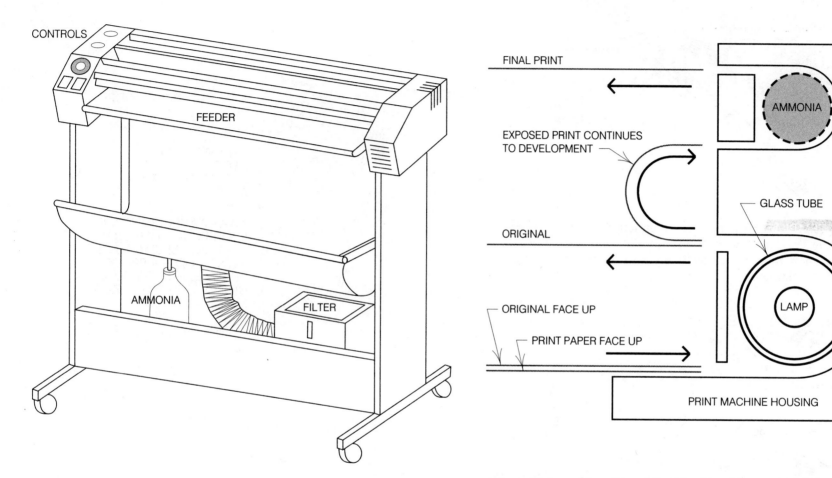

**PRINT MACHINE:** The most common method of reproduction uses a whiteprinter. In this process a positive image is formed with either blue or black lines on a white background. It is known as a *diazo* print. The term "blueprint," though still used today, really refers to the older *ozalid* process where the image is negative. This process produces white lines on a blue background.

Both prints described above are produced on specially treated paper. Another type of print that can be produced with this machine is called a sepia. A *sepia* is a print on special media that resembles vellum. The result is a print that can serve as another original. More drawing can be done on it and prints can be generated from it.

In a print machine, the original is placed face up on top of a piece of light sensitive paper. When fed into the printer, the lamp shines through the original and is blocked where lines are drawn. The original and print come out of the printer where they are separated. The sensitized material is fed through an upper chamber where ammonia gas fixes the remaining light sensitive material to make the final print.

**ENGINEERING COPIER** (not shown): This machine looks the same as a floor model office copier except that it is built to handle large size documents. It can copy on any type of media and can reduce and enlarge. It is a very versatile piece of equipment for manipulating originals or creating new ones.

# CAD Drafting Equipment

Computers and their accessories have emerged in recent years as wonderful new tools to assist architects, engineers, and contractors in bringing more and better services to their clients.

There is a wealth of software available to help in all phases of the construction process, but computer aided drafting or computer aided design (**CAD**) programs have perhaps made the biggest impact.

Originally, design professionals saw CAD as a way of producing working drawings faster. This proved to be the case, but users saw an even greater benefit: the ability to manipulate the data once it is input. A file, once built, can easily be transferred to a consultant; parts of it can be edited out and used elsewhere; the scale can be changed at will; and numerous originals can be produced.

Today this flexibility has reached new heights. While manual drafting is essentially two dimensional, projects are produced in CAD programs in three dimensions. From one file many two-dimensional drawings can be generated, movies taking the viewer around and through a building are possible, and photo-realistic renderings can be created. Even cost estimates and lists of materials can come from this single file.

Many professionals believe that manual drafting will never completely disappear from the construction industry. However most believe that computer aided drafting will continue to encroach on manual drafting.

Regardless, a beginning drafter must see CAD for what it is, another tool to help communicate designs among those in the construction industry and their clients. Whether a hand or a machine actually draws a line, it is the cognitive skills behind that line that are important.

Most CAD systems today are connected to a **network**, a system of interconnected computers and other devices allowing sharing of resources. Assistive devices known as **peripherals** offer a variety of ways to input, output, store, and retrieve data.

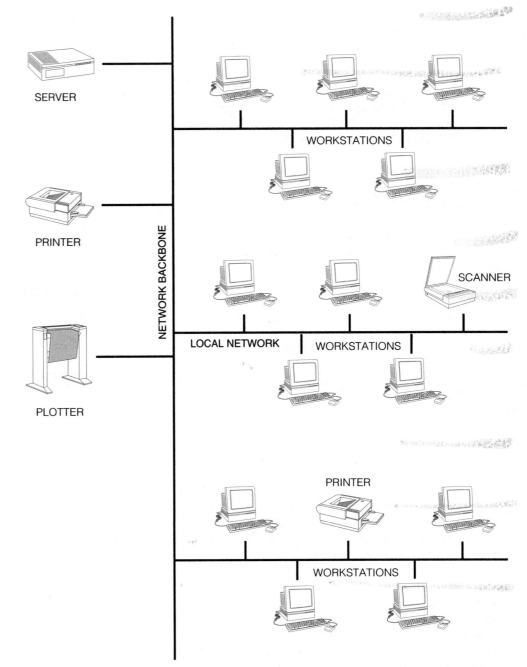

*Example configuration of a network.*

WORKSTATION: Workstation is the name given to all of the components listed below; i.e., all the necessary equipment required for one worker to draft electronically.

CENTRAL PROCESSING UNIT (CPU): The CPU is the brain of the computer system, a box containing the electronics, which do all of the processing required to operate the system. The CPU also contains a hard disk drive, which stores data and programs.

HARD DISK DRIVE: This is a device usually internal to the CPU that stores applications and data. It holds much more data than the diskettes described below.

MONITOR: The monitor is the visual display system, which transforms electronic data from the CPU into graphics on a screen.

KEYBOARD: The keyboard is an alpha numeric input device used to send commands or data to the CPU.

MOUSE: A mouse is an input device used to select commands on the monitor. A similar looking device called a *puck*, has cross hairs and is used with a menu or digitizing tablet. An electronic pencil, called a *stylus*, works similarly to a puck.

MENU TEMPLATE: This is a menu overlay on a digitizer, which allows the user to send commands to the CPU by simply selecting a cell on the menu.

DIGITIZING TABLET: The digitizing tablet is an input device used to send data to the CPU by electronically tracing an existing drawing or by using a menu tablet. Digitizing tablets can be as large as a drafting table.

DISKETTE OR FLOPPY DISK: A storage device for small amounts of data. It is a small, circular piece of magnetic medium encased in a plastic jacket. The most common sizes are 3-1/2" and 5-1/4". Data can be accessed or saved when the diskette is inserted into the floppy disk drive of the CPU.

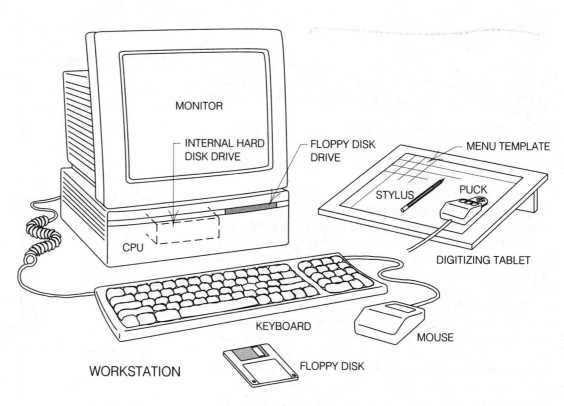

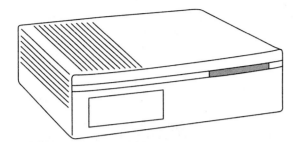

SERVER: A server is an electronic storage device where large amounts of data can be stored and retrieved when needed. It is actually a computer with a very large hard disk.

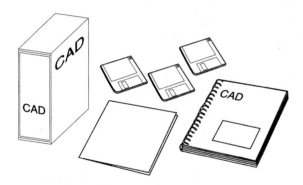

SOFTWARE: Hardware cannot do anything without software (programs) to drive it. *System software* allows the various devices to "communicate" with each other. Common system software includes MS-DOS®, Macintosh®, OS-2®, and UNIX®. For the most part workstations are built to run one of these programs. Together they are called an *operating system* or *platform*.

Application software programs perform specific tasks such as drafting, word processing, cost estimating, etc. Popular CAD application programs include ArchiCAD®, Arris®, AutoCAD®, Microstation®, and a variety of third party programs that add additional functions.

SCANNER: A scanner is an input device that reads a document (drawing, text, photograph) and converts it to electronic data (a file).

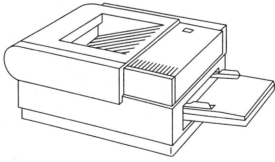

PRINTER: This is an output device where the image is transferred to the media by using a series of tiny dots (dot matrix) rather than true lines. This is done by impacting a ribbon or with toner as in a copy machine. When laser technology is used, the dots are much closer together and the image is sharper. This is known as a laser printer. When the dots are placed as tiny droplets of ink, this is known as ink jet printing. Clarity ranges from 150 to 1200 dots per inch (dpi).

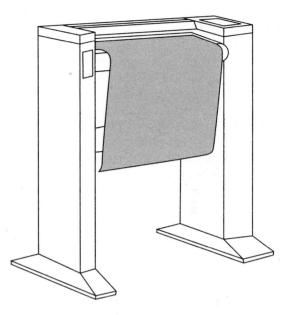

PLOTTER: A plotter is an output device that converts electronic data from the CPU into lines (vectors) on drafting media. The roll plotter variety stands vertically since both the pens (or lead) and media move. In the flat bed variety the media remains stationary while only the pens or (lead) move. A thermal or electrostatic plotter works like a large scale laser printer, fusing toner to the media with heat.

# TECHNIQUES

These techniques range from tips on how to secure the media on the drafting board to methods of sharpening the lead on a compass. This chapter covers the basics and is, by no means, a complete collection of drafting techniques. Drafters continue to learn new techniques all the time and may even develop new ones of their own.

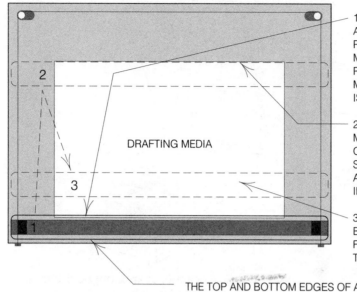

DRAFTING MEDIA

1. WITH THE PARALLEL BAR (OR T-SQUARE) AT THE BOTTOM OF THE DRAFTING BOARD, PLACE THE BOTTOM EDGE OF THE DRAFTING MEDIA JUST BELOW THE TOP EDGE OF THE PARALLEL BAR. THIS WILL PREVENT THE MEDIA FROM BEING RIPPED WHEN THE BAR IS PUSHED TOWARDS THE TOP OF THE BOARD.

2. NOW, HOLDING THE MEDIA IN PLACE, MOVE THE PARALLEL BAR TO THE TOP EDGE OF THE MEDIA AND ALIGN THEIR EDGES. SLIDE THE PARALLEL BAR DOWN SLIGHTLY AND TAPE THE TOP CORNERS OF THE MEDIA IN PLACE AS SHOWN BELOW.

3. SLIDE THE PARALLEL BAR DOWN ALMOST BACK TO THE BOTTOM, STRETCHING AND FLATTENING THE MEDIA AS YOU GO. TAPE THE BOTTOM TWO CORNERS IN PLACE.

THE TOP AND BOTTOM EDGES OF A T-SQUARE OR PARALLEL BAR ARE PROBABLY NOT PARALLEL. DRAFTERS SHOULD THEREFORE NEVER USE THE BOTTOM EDGE TO DRAW WITH OR SET EQUIPMENT OR MEDIA AGAINST.

## Securing the Media

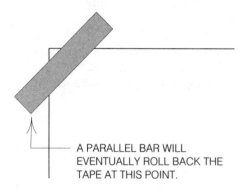

A PARALLEL BAR WILL EVENTUALLY ROLL BACK THE TAPE AT THIS POINT.

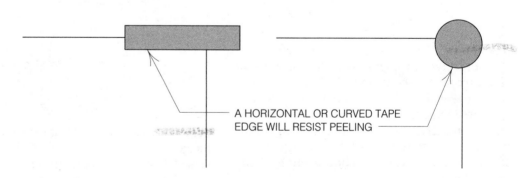

A HORIZONTAL OR CURVED TAPE EDGE WILL RESIST PEELING

When attaching media to the board it may seem that a piece of tape placed at an angle is the most secure method. However the parallel bar or T-square will "catch" the corners of the tape as it is moved up and down. Within a short time the tape will be rolled off, and the media will move on the board.

Tape placed horizontally or drafting dots placed at the corners will hold the media securely yet resist the rolling effect of the parallel bar or T-square. The media should be stretched slightly as it is being taped down so as to avoid ripples or looseness, which can effect drafting quality.

## Cleanliness

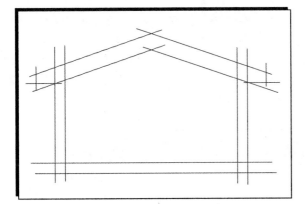 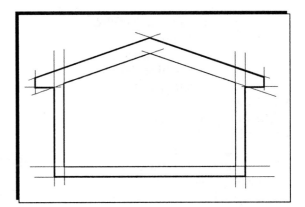

Keeping a drawing clean is important. The first method of accomplishing this is to draw lines in three stages. First, draw layout lines. These are light lines drawn with a hard lead such as 4H, which will not smear easily.

Next draw final lines ranging from very thin to medium thickness. These may then be drawn over the layout lines. These lines are produced using medium hardness leads and will only smear to a small degree. Layout lines need not be erased.

Draw thick and very thick lines last because they are made with soft leads or thick pen points, which smear very easily.

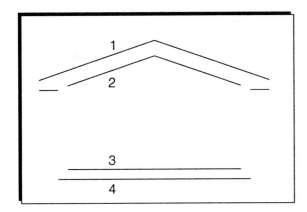 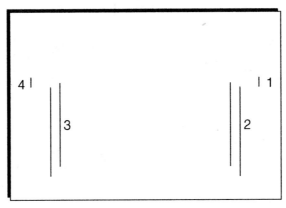

Another technique for keeping drawings clean is to avoid dragging equipment over lines that have already been drawn. To do this it is generally a good idea to work from the top down and from right to left (left to right for left handers). The numbers shown above indicate the order in which the lines should be drawn.

Other techniques can be used to keep drawings clean. For example, when a sheet contains more than one drawing, tape a piece of tracing paper over the part of the drawing you want to protect. In that way the paper collects the dirt and can then be disposed of.

A drafter's hands become dirty and cause as much spreading of dirt as equipment. Occasional hand washing can really contribute to a clean, smudge-free job.

## Accuracy

Besides line work, accuracy is also one of the key components of good drafting. When measuring before drawing a line, it is important to measure exactly along the axis of the line.

In other words, if measuring for a horizontal line, the scale should be placed perfectly horizontal by resting it on the parallel bar. If measuring for a 45 degree line, it should be placed on a 45 degree triangle, and so forth.

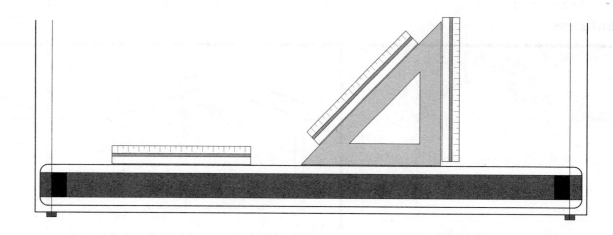

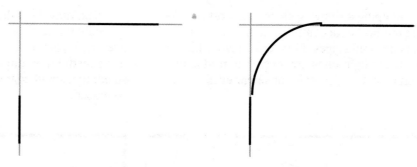

When drawing a curve that is tangent to a line or lines, draw the curve first and then the lines. If the lines are drawn first the result can look like the drawing above.

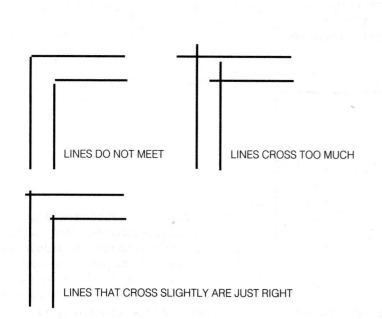

LINES DO NOT MEET

LINES CROSS TOO MUCH

LINES THAT CROSS SLIGHTLY ARE JUST RIGHT

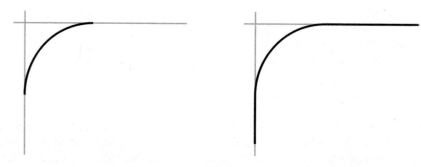

The way lines meet is also a measure of accuracy. When two straight lines meet they should overlap slightly so that the point of intersection is obvious.

If the curve is drawn first it is much easier to make a perfect joint because the straight lines can begin exactly at the ends of the curves.

## Angles and Curves

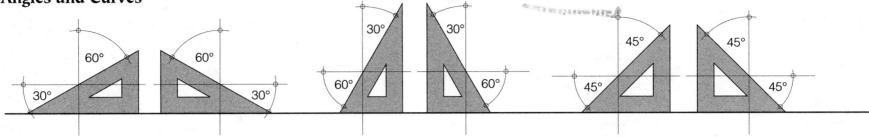

Most lines can be drawn using the 30, 45, 60, and 90 degree angles found on standard triangles.

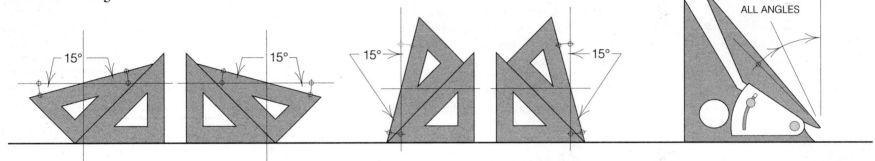

Using both standard drafting triangles together, a drafter can draw angles at 15 degree increments. For lines at other angles an adjustable triangle is the tool of choice. It has markings down to half a degree, certainly accurate enough for most drawings.

Except for circles and ellipses, which can be drawn with templates, irregular curves need to be drawn with French or flexible curves. In many situations a curve is a compound one, which requires the drafter to position and trace the French curve at more than one location. Connection of one segment to the next should be seamless.

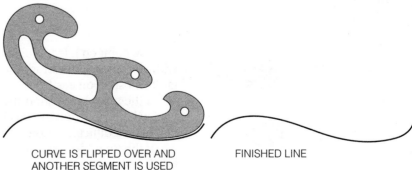

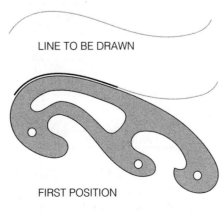

LINE TO BE DRAWN

CURVE IS FLIPPED OVER AND
ANOTHER SEGMENT IS USED

FINISHED LINE

FIRST POSITION

## Lead and Ink

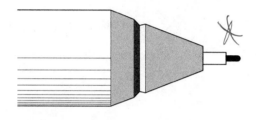

Standard wooden lead pencils are not generally used for drafting because the drafter cannot control the point very well when sharpened in a standard pencil sharpener. However, drafters do use hard lead pencils for layout lines and lettering guidelines.

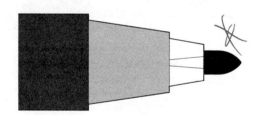

Mechanical pencils hold very thin leads that do not need to be sharpened. They really only produce one thickness of line so they are not suitable for all line work. They are handy for lettering and for producing lines of medium thickness.

A lead holder gives a drafter the most flexibility. By manipulating the point using a lead pointer, the drafter can produce lines of any thickness.

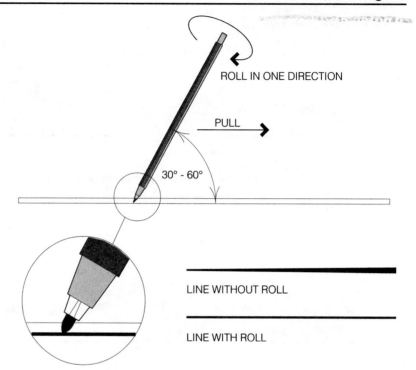

ROLL IN ONE DIRECTION

PULL

30° - 60°

LINE WITHOUT ROLL

LINE WITH ROLL

The secret to producing consistent lines and keeping the point sharp is to roll the lead holder while pulling it across the parallel bar or triangle. If the lead holder is not rolled the line produced will get progressively wider.

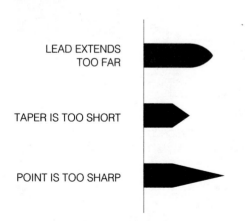

LEAD EXTENDS TOO FAR

TAPER IS TOO SHORT

POINT IS TOO SHARP

The point on a lead holder is critical. If the lead protrudes too far it will snap off easily.

If the taper is too short the lead will become dull too quickly and will be difficult to control.

If the tip is too sharp it too will break easily. The best shape, as shown in the lead holder above, has a long taper and slightly rounded point.

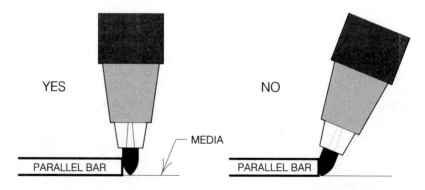

YES

NO

MEDIA

PARALLEL BAR

PARALLEL BAR

Lead holders should be perpendicular to the edge of the equipment so that lead does not get ground into the edge of the equipment where it can later loosen and smudge the drawing.

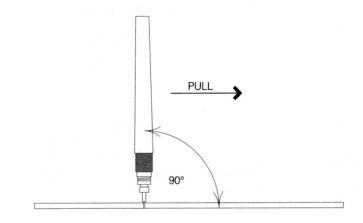

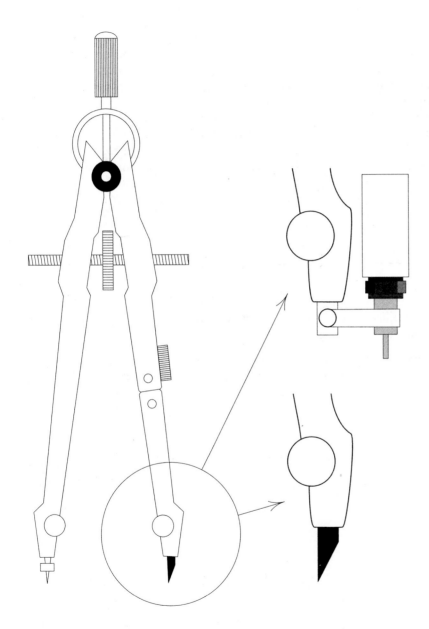

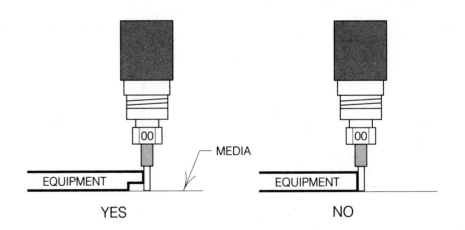

Technical pens dispense ink best when held vertically. Like lead holders, a pulling motion across the media provides the best results. Light pressure is all that is needed for technical pens.

YES          NO

If a technical pen is used with a square edged triangle or parallel bar the ink will run under the equipment and create a smear.

A good parallel bar or offset edge triangle creates a gap between the pen point and the point of contact, eliminating the problem of bleeding under the instrument.

Adhesive backed pads, coins, or even droplets of glue may be attached to equipment to raise it for inking purposes.

Lead for a compass is sharpened by sanding it to a flat point. If ink is used in a compass, a pen point can be mounted in place of the lead with a standard fitting.

# KEY TERMS

drafting board
surface cover
T-square
drafting machine
parallel rule
drafting media
bumwad
onion skin
tracing paper

vellum
drafting film
Mylar®
tooth
drafting tape
drafting dots
lead holder
mechanical pencil
drafting leads
lead pointer
technical pen
ink

architect's scale
engineer's scale
metric scale
adjustable triangle
30-60-90 triangle
45-45-90 triangle
French curve
flexible curve
template
bow compass
friction compass
beam compass

eraser
drafting brush
erasing shield
drafting powder
lettering template
Leroy® lettering
lettering machine
transfer lettering
burnish
print machine
diazo
ozalid

sepia
engineering copier
CAD
network
peripheral
workstation
CPU
hard disk drive
monitor
keyboard
mouse
puck

stylus
menu template
digitizing tablet
floppy disk
server
system software
operating system
platform
application software
scanner
printer
plotter

# PRACTICE EXERCISES

1. Go back through the chapter and review all of the illustrations. Cover up the descriptions and practice naming each item and explaining its purpose.

2. Make the drawings below, first with lead on vellum, then with ink on film. Sheet size is 8-1/2" x 11"; borders are 1/2" bottom and right, 3/4" top and left; line spacing is 1/4".

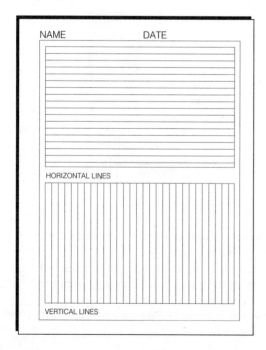

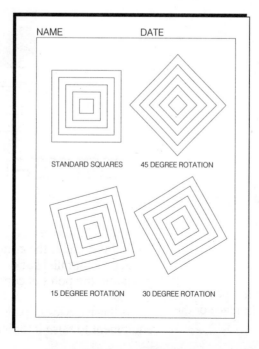

# 3

# LINE WORK

BREAK LINES

PHANTOM LINES

PROFILE LINE

GRADE LINE

## OBJECTIVES

*By the end of this chapter students should be able to:*

- Given a line type, identify the weight (thickness) at which it is drawn.
- Identify and draw all the basic line types.
- Draw five distinct line thicknesses with ink or lead that are black and consistent.
- Draw various lines that meet or intersect to standards established in this chapter.

# LINES

Lines are the backbone of construction drawings. They are not all the same, however. Some are literal representations of objects, some simulate reality, and others are abstractions. A closer look at each is important.

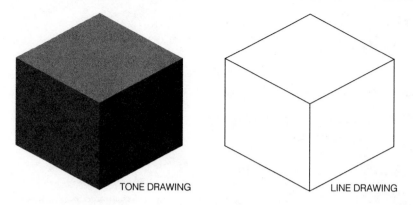

TONE DRAWING            LINE DRAWING

The most common use of lines on a drawing is to define an object by representing its edges as lines. This is not how an object really appears. In reality, the edges of an object appear to be lines but they have no thickness. Edges are defined by the difference in tone from one surface to another caused by a given lighting condition. Thus the tone drawing above is closer to reality, but the line drawing next to it is a much more efficient way of representing the object.

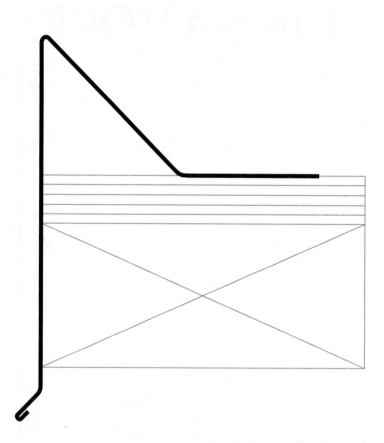

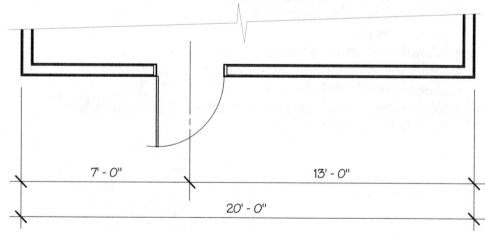

Lines can be literal representations of objects. The line above is exactly how a piece of sheet metal would appear looking at its edge. This is the simplest type of line, but not the most common on a given drawing.

Lines can also represent things we cannot see. The door shown above is certainly a real object, but its path of travel (shown as an arc) is represented with a thin line.

The dimension lines shown at the bottom of the illustration show another use of lines called **abstractions**. There are certainly no such lines on the floor of the room, but the drafter adds them to present abstract dimensional information.

Essentially there are two types of lines a drafter uses on a drawing:
- those meant to be seen
- those not meant to be seen

Those *not* meant to be seen are lines that the drafter uses to set up the drawing before applying the final line work. These lines are drawn with lead and with very little pressure so that they appear gray. They may be visible on the original drawing, but usually will not show up on any type of reproduction. The two types are explained at right.

Lines that are meant to be seen are obviously those that comprise the actual final drawing. They are in most cases drawn (at least partially) right over the lines described above. Although they may vary in thickness and configuration, the characteristic they all share is that they are black, dense, and opaque in order to make high quality reproductions.

*Lines not meant to be seen are for the drafter's use only. They are drawn dark enough to be seen on the original but light enough not to show up on a reproduction.*

*Final lines, those meant to be seen, vary in thickness and character, but all are dense, black, and opaque. There are over a dozen types.*

# LAYOUT AND GUIDELINES

*Layout lines* are drawn on the original by the drafter to establish the basic shape of the object to make sure the location is appropriate before proceeding. They should be drawn just dark enough so that the drafter can see them. A 4H lead or harder is a good choice for producing them. Not every final line needs to be drawn first as a layout line; in fact, the majority of lines can be drawn directly as final lines.

*Guidelines* are roughly to lettering what layout lines are to objects. They serve as visual controls to help the drafter keep lettering horizontal and of consistent height. Their darkness should be exactly the same as layout lines.

It is not critical that layout lines and guidelines be invisible on reproductions. It is generally acceptable if these lines are slightly perceptible.

Many drafters use non-photo blue pencils for layout lines. The color makes them easy to see, but care should be taken not to apply the lead too thickly. The blue lead is waxy in nature and could make applying the final line difficult.

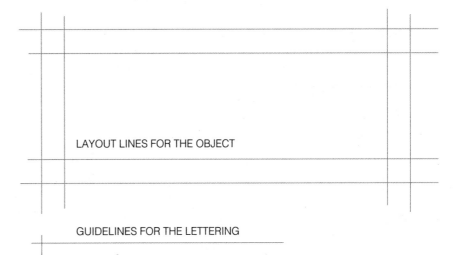

LAYOUT LINES FOR THE OBJECT

GUIDELINES FOR THE LETTERING

*Layout lines are used here to define the outline of the rectangular shape shown. Later the drafter would draw over the layout lines to complete the object. Guidelines are similar in that they block out the area where text will be placed. They differ from layout lines in that the lettering itself is not outlined.*

# FINAL LINES

## Object Lines

*Object lines* are probably the most important lines on a drawing because they define what a drafter is depicting. In most cases object lines represent an edge; the remainder of times they represent where two objects meet on the same plane. Object lines are continuous, and most importantly, can and should be drawn to various thicknesses.

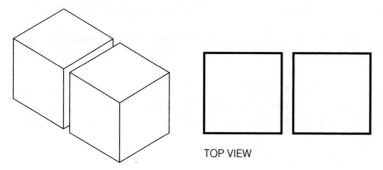

TOP VIEW

When objects are drawn, the edges may be shown as object lines of medium thickness. The medium thickness helps to give the illusion that there is depth from the top of the objects to the surface on which they sit.

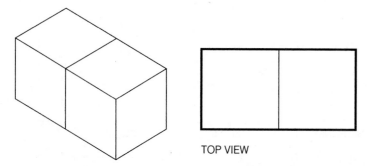

TOP VIEW

If two objects are in contact with each other, one edge of each becomes a joint. As such it no longer warrants a medium line but is drawn thin to show that there is no change of depth. A medium line should be drawn around the perimeter of the two objects because a change of height does occur there.

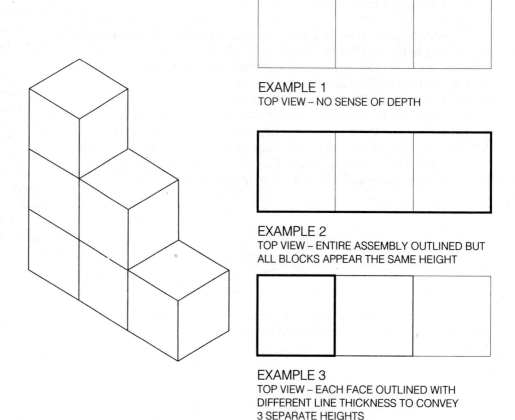

**EXAMPLE 1**
TOP VIEW – NO SENSE OF DEPTH

**EXAMPLE 2**
TOP VIEW – ENTIRE ASSEMBLY OUTLINED BUT ALL BLOCKS APPEAR THE SAME HEIGHT

**EXAMPLE 3**
TOP VIEW – EACH FACE OUTLINED WITH DIFFERENT LINE THICKNESS TO CONVEY 3 SEPARATE HEIGHTS

By varying the height of the objects, another principle may be seen. The drafter must now give the illusion of depth at three levels. If all the lines are drawn with the same thickness as in Example 1, no sense of depth is conveyed.

Based on the top view drawing of two cubes pushed together at left, one might initially think that outlining the assembly and drawing the two center lines thinner would help define the assembly. However, this too fails to portray depth as shown by Example 2 above.

Only by treating each face as a separate plane can we give an adequate sense of depth. This is shown in Example 3. The face closest to the viewer is drawn with a medium line, the next face down is drawn as a thin line, and the face of the final cube is drawn as a very thin line.

## Hidden Lines

*Hidden lines* are object lines that cannot be seen because they are either covered by an object or they face away from the viewer. Hidden lines are produced by alternating line segments and spaces. The sizes of the line segments and spaces are not as important as their proportions. Generally, the lines should be about twice the length of the spaces. On large drawings that may mean 1/4" lines with 1/8" spaces, whereas on small scale drawings, 1/8" lines with 1/16" spaces could be used.

Drafters should not try to measure hidden lines when drawing since this would be a waste of time. One needs to develop an eye for the spacing, which can only come with practice. On the other hand, most CAD packages allow the operator to set line parameters so the desired result is assured.

LINE SEGMENTS TWICE AS LONG AS SPACES

Generally, the thickness of hidden lines is medium. Some drafters vary the thickness depending on the object lines surrounding the hidden lines, i.e., the thickness of the hidden lines can be one step thinner than the surrounding object lines. This alternate method is illustrated at right.

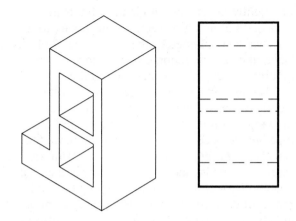

*The holes in this block are shown on the drawing of the back face as thin lines since the block is outlined with medium lines.*

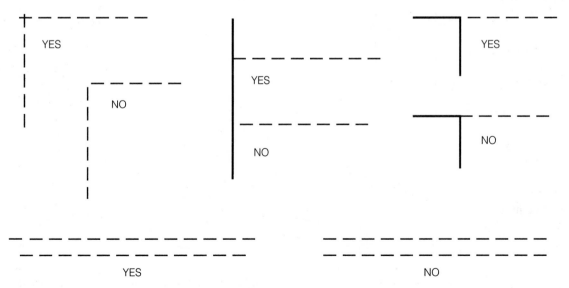

*Hidden lines should be drawn according to certain conventions as shown above. The last convention is not a hard and fast rule because it is not easy to accomplish using CAD.*

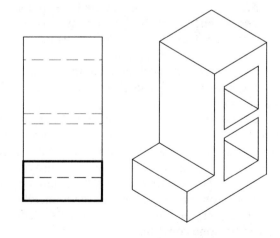

*In this case the forward most face of the block is outlined with medium lines, but the face set back is outlined with thin lines. The hidden lines representing the holes are therefore drawn as thin and very thin lines respectively.*

## Section, Material Indication, and Center Lines

Frequently to get a better idea of what is inside a completely enclosed but hollow object, it is necessary to draw a section of it. This will be explained further in Chapter 8 of this text, but basically, to create a section drawing make an imaginary cut through an object to expose the interior and then draw it.

The example on this page shows a column, which is more clearly explained when the front view is accompanied by a section drawing. In order to tie the front view and the section drawings together we need to show exactly where we took our imaginary cut. This is the purpose of a **cutting plane line** (also called a **section line**). The line is made by alternating one long and two short dashes. The rules for hidden line lengths can be applied here for the short dashes; the long dashes can be varied such that a few but at least one grouping of small dashes occurs over the entire length.

The cutting plane line should go entirely across the object being cut. The ends need to be "bent" at right angles to indicate the direction of view. The symbols used at the ends vary and will be covered later. Cutting plane lines should be very thick so that they are easy to pick out.

**Material indication lines** are lines that show what the object is made of. They may also be referred to as **symbol lines**. They are most frequently used to show materials in section but can be used to render a surface as well. Some material indication lines look like the actual material; most are mere representations and even abstractions. Symbol lines should be drawn very thin since they are support information.

**Center lines** indicate the center(s) of symmetrical objects. This can occur in more than one axis as the sample on this page illustrates. All the rules that apply to cutting plane lines apply to center lines except that center lines require only one dash instead of two and they should be thin rather than very thick.

Notice on the section that at the point where the two center lines intersect, the dashes intersect. This is common practice and should be done consistently.

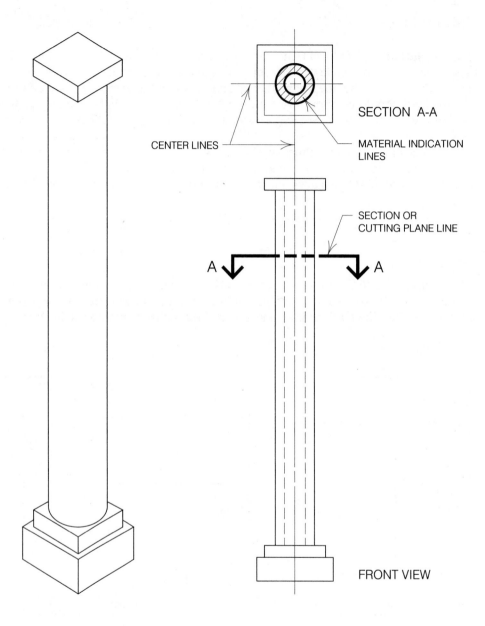

*Center lines shown here mark the horizontal and vertical axes of the column in both views. Hidden lines on the front view indicate the location of the inside surface of the column. The section line indicates where the cut is taken to produce Section A-A.*

## Extension and Dimension Lines

The illustration at the right shows the use of extension, dimension, leader lines, and terminators. All of these are found together on a drawing and comprise the system of dimensioning.

*Extension lines* are those that project out from the object. They are solid lines of thin weight. They should be drawn a noticeable distance from the object (about 1/16") and extend far enough away to avoid confusion with other lines.

*Dimension lines* are those that are perpendicular to extension lines and thus parallel to the object. The points at which they intersect the extension lines make them exactly the same length as the face of the object they are parallel to. Thus we write the dimension of the line and the corresponding surface at the dimension line. For AEC drawings the dimension is usually placed above the line. Like extension lines, dimension lines should be thin.

## Leader Lines and Terminators

*Leader lines* are those that indicate where a dimension or string of text belongs but doesn't fit. They begin at the text or dimension and end at the appropriate location with a terminator. They may be constructed using continuous straight lines or arcs and should be of thin line weight.

*Terminators* are symbols drawn at the intersection of extension and dimension lines to indicate precisely where the dimensions begin and end. They are also used at the end of leader lines. There are several types of terminators a drafter can choose from. Their thickness ranges from thin to medium.

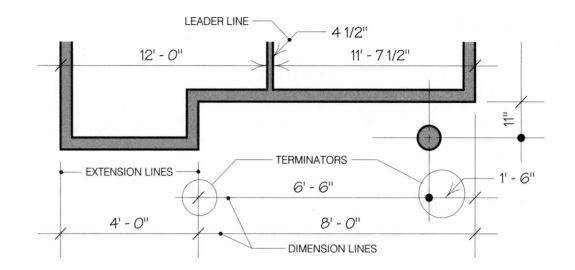

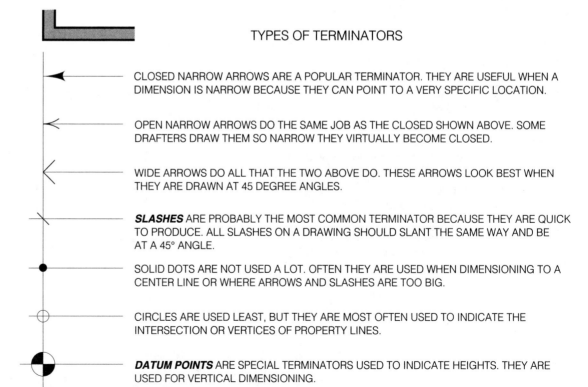

TYPES OF TERMINATORS

CLOSED NARROW ARROWS ARE A POPULAR TERMINATOR. THEY ARE USEFUL WHEN A DIMENSION IS NARROW BECAUSE THEY CAN POINT TO A VERY SPECIFIC LOCATION.

OPEN NARROW ARROWS DO THE SAME JOB AS THE CLOSED SHOWN ABOVE. SOME DRAFTERS DRAW THEM SO NARROW THEY VIRTUALLY BECOME CLOSED.

WIDE ARROWS DO ALL THAT THE TWO ABOVE DO. THESE ARROWS LOOK BEST WHEN THEY ARE DRAWN AT 45 DEGREE ANGLES.

*SLASHES* ARE PROBABLY THE MOST COMMON TERMINATOR BECAUSE THEY ARE QUICK TO PRODUCE. ALL SLASHES ON A DRAWING SHOULD SLANT THE SAME WAY AND BE AT A 45° ANGLE.

SOLID DOTS ARE NOT USED A LOT. OFTEN THEY ARE USED WHEN DIMENSIONING TO A CENTER LINE OR WHERE ARROWS AND SLASHES ARE TOO BIG.

CIRCLES ARE USED LEAST, BUT THEY ARE MOST OFTEN USED TO INDICATE THE INTERSECTION OR VERTICES OF PROPERTY LINES.

*DATUM POINTS* ARE SPECIAL TERMINATORS USED TO INDICATE HEIGHTS. THEY ARE USED FOR VERTICAL DIMENSIONING.

## Break, Phantom, Profile, Grade, and Border Lines

You will notice on the illustration on this page that there is not enough room to show the entire object. To indicate this, break lines have been used. *Break lines* are continuous, very thin lines with jagged interruptions in them one or more times over their length. Their length is slightly more than that of the object they break, and they are usually drawn at a slight angle to connote the concept of a break.

*Phantom lines* are those that represent something other than a real object or the outline of an object that has been removed for clarity. They can represent something we cannot see such as an indication of movement. The door swing in both views at right is a phantom line.

Phantom lines in mechanical drawings are double dash lines. In the AEC world they can take any form, but are usually shown as a continuous or hidden line as in the illustration at right. Since these lines represent non-visible or imaginary objects, they should be very thin.

*Profile lines* are nothing more than object lines, but as illustrated earlier in the chapter, they are a way of emphasizing the outline of an assembly. They are of thick line weight.

*Grade lines* are also object lines but represent where an object meets the soil on a construction project. Because this is a critical point in a drawing and a project, the line is drawn very thick.

*Border lines* are those used around the edge of a sheet to define the drawing area. They may also be used to sub-divide the drawing area into smaller parts. They are drawn very thick. The summary of lines on the facing page is surrounded and divided with border lines.

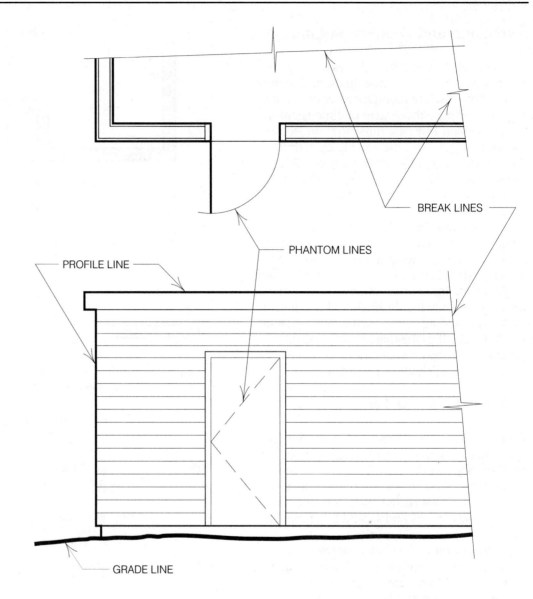

## LINE SUMMARY

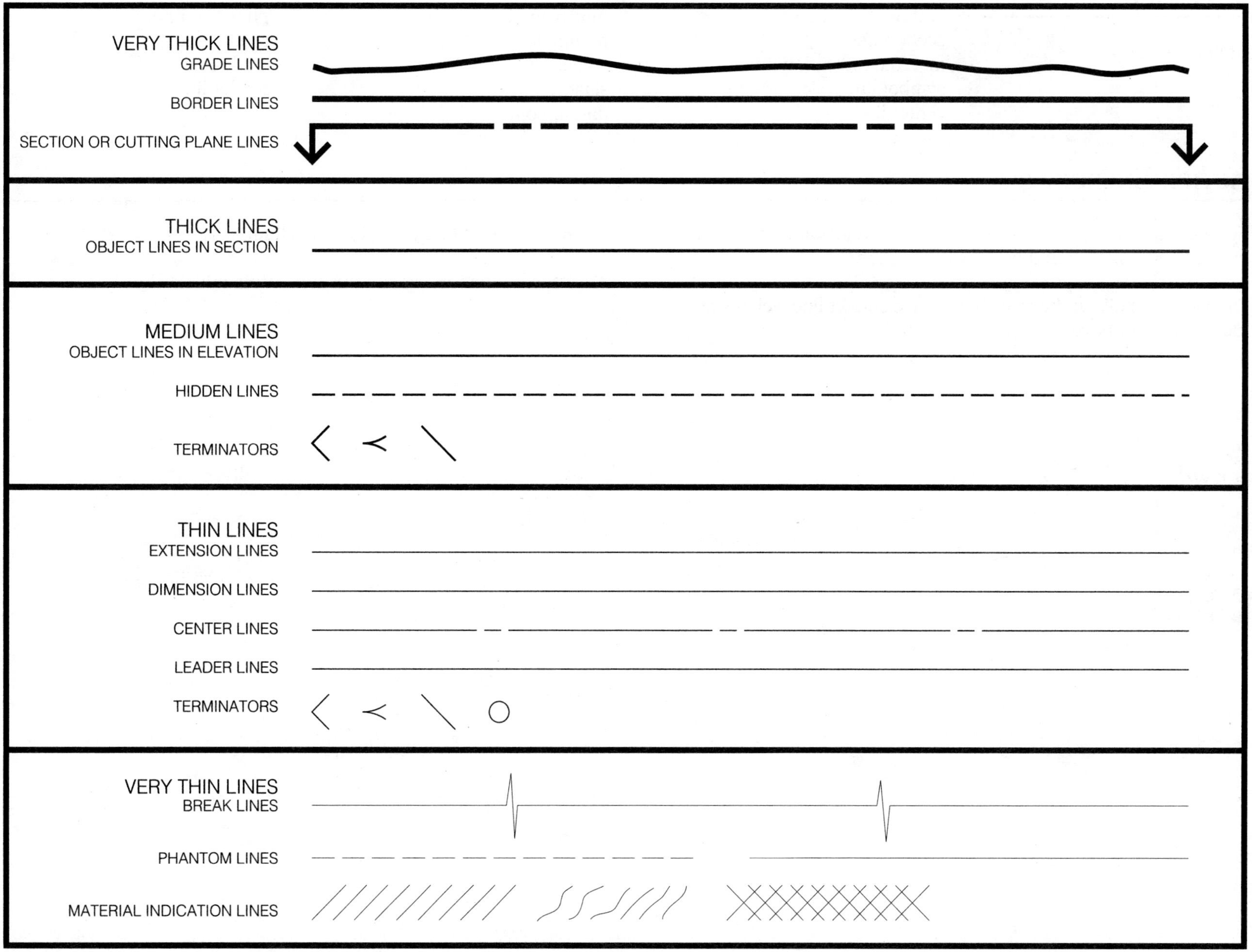

**VERY THICK LINES**
GRADE LINES

BORDER LINES

SECTION OR CUTTING PLANE LINES

**THICK LINES**
OBJECT LINES IN SECTION

**MEDIUM LINES**
OBJECT LINES IN ELEVATION

HIDDEN LINES

TERMINATORS

**THIN LINES**
EXTENSION LINES

DIMENSION LINES

CENTER LINES

LEADER LINES

TERMINATORS

**VERY THIN LINES**
BREAK LINES

PHANTOM LINES

MATERIAL INDICATION LINES

# KEY TERMS

layout line
guideline
object line

hidden line
cutting plane line
section line
material indication line
symbol line
center line

extension line
dimension line
terminator
leader line
slash
datum point

break line
phantom line
profile line
grade line
border line

# PRACTICE EXERCISES

1.  For the drawings below, identify (name) each line and indicate at which thickness it should be drawn.

2.  Draw both details at the size shown. Use correct line weights as identified in 1 above. Use lead on vellum.

3.  Repeat number 2 using ink on vellum.

4.  Repeat number 2 again, this time using ink on film.

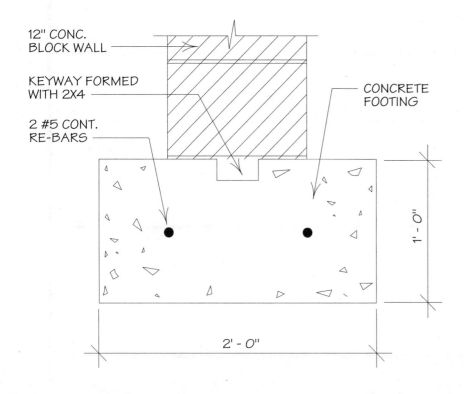

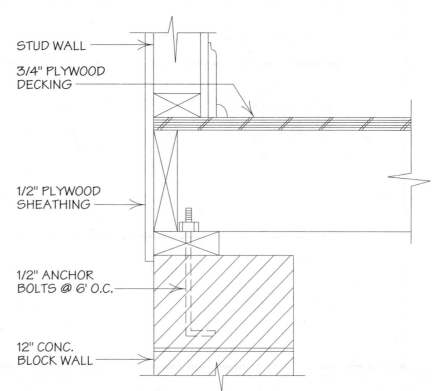

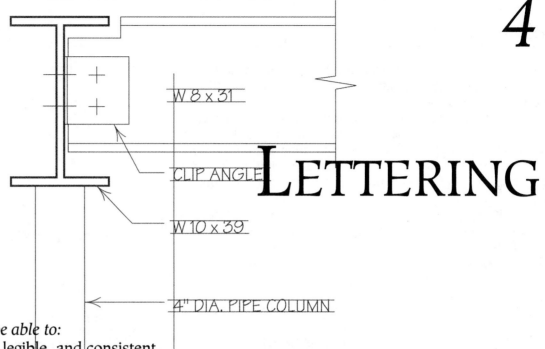

# 4

# LETTERING

## OBJECTIVES

*By the end of this chapter students should be able to:*
- Produce hand lettering that is sharp, legible, and consistent.
- Space letters by equal areas and space words from 1 to 1.5 letter widths.
- Produce lettering to four different heights with proper thickness.
- Recognize and be able to place lettering at an appropriate angle and in the appropriate place on a drawing.

# LETTERING STYLE

Although the graphic part of a drawing provides the most critical information, written communication is almost always needed to make the sheet completely understandable.

An important objective of lettering found on a drawing is to support the graphic information as efficiently and as clearly as possible. It can be said that it must communicate without complicating. For this reason the majority of construction related firms that make drawings prefer that the lettering be simple, easy to read, and not impart any style onto the drawings.

A simple lettering style also makes it easier for many drafters working on a project to produce a set of drawings that are cohesive in the way they look. This is why many firms use mechanical lettering devices (such as Leroy lettering sets) or computer aided drafting. With these tools the end result is consistent-looking drawings despite the fact that many people have worked on them.

Professional quality lettering has the following characteristics:
- only **uppercase** letters (capitals)
- letters should be vertical (not slanted or **italic**)
- letters should touch guidelines
- the width of full size letters should be 75% - 100% of their height
- each letter should be made the same each time

These characteristics are illustrated below and on the facing page.

*The first step in producing good quality lettering is to establish guidelines. These will help the drafter keep the lines of lettering horizontal.*

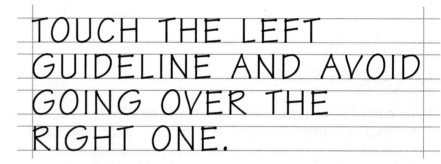

*Top and bottom guidelines are certainly critical, but left and right ones are equally important. Each line of lettering should begin by touching the left guideline, and no letter should extend beyond the right one.*

ABCDEFGHIJKLMNOPQRSTUVWXYZ1234567890
ABCDEFGHIJKLMNOPQRSTUVWXYZ 1234567890
ABCDEFGHIJKLMNOPQRSTUVWXYZ1234567890
ABCDEFGHIJKLMNOPQRSTUVWXYZ1234567890

*The letters shown above represent professional quality lettering shapes. Beginning drafters may model their letters after these but in time will develop their own style. Using lead with a chisel point (example at the top) produces narrow vertical lines and wide horizontal ones, giving a strong character. Letters that are slightly wider at the bottom than at the top also have a more pleasing, stable look to them. In all cases, the letters are shaped carefully; speed will come with experience.*

A CLEAN SIMPLE STYLE IS BEST

THE HAND LETTERING OF SEVERAL DRAFTERS
WILL SHOW SMALL BUT INTERESTING VARIATIONS
BETWEEN INDIVIDUALS.

GOOD LETTERING TELLS PEOPLE YOU ARE A
SKILLED TECHNICIAN, SHOW YOUR SKILL
WHENEVER YOU LETTER,

GOOD FREEHAND LETTERING SKILLS MAY
BE THE KEY TO YOUR FIRST EMPLOYMENT
OPPORTUNITY

HAND LETTERING CAN EVEN BE TRANSFORMED
INTO A COMPUTER FONT. THIS ONE WAS CREATED
BASED ON AN ARCHITECT'S ACTUAL LETTERS.

*Several good lettering styles are shown above. Although noticeably different, there are certain characteristics that they all exhibit: They are all simply shaped and thus quick and easy to produce; they are all easy to read; and they all look good together.*

## LETTERING SIZE AND THICKNESS

Although there are no strict rules for size, the following guidelines may be helpful to follow:

LETTERING HEIGHT

| manual | CAD | where used |
|---|---|---|
| >1/4" | > 18 pt. | title blocks |
| 1/4"-3/16" | 14-18 pts. | drawing titles |
| 3/16"-1/8" | 12-14 pts. | subtitles or parts of a main drawing |
| 1/8"-3/32" | 10-12 pts. | general notes and dimensions |
| 3/32" | 8-10 pts. | miscellaneous notes |

The guidelines for CAD are given in **point sizes** (abbreviated "pts."). Many CAD packages size lettering with this system. One inch equals 72 points, which is a system of measuring used in the publishing industry. In other CAD packages the drafter draws at full scale. In this case the lettering needs to be scaled up to full size (sized with the final output scale in mind).

There will certainly be times when the correct size lettering will not fit in the space available. Generally it is better to break the rules and use the next smaller lettering than to force the larger lettering into the space and make it unreadable.

The specific lettering heights suggested are also not sacred. More important is the relationship between the heights selected. If using the metric system, for example, the drafter might select 2, 4, 6, and 8 mm rather than fractional inches or point sizes.

Certain organizations have minimum lettering sizes too. There may be office standards or government standards for municipal work. Whenever drawings will ultimately be reduced to microfilm for final storage, for example, the minimum suggested lettering size is 1/8".

Title Blocks

DRAWING TITLES

SUBTITLES

GENERAL NOTES AND DIMENSIONS

MISCELLANEOUS NOTES

LETTERING THICKNESS

*Lettering larger than 1/4" should not be standard hand lettering. It should be done by outlining typeset letters by hand with a thick line as at left. This is a common technique used for project titles found in the title block. The alternative is machine-made letters as shown on the facing page.*

*Drawing titles at 1/4" with a thick line*

*Subtitles at 3/16" with a medium line*

*General notes at 1/8" with a thin line*

*Miscellaneous notes at 3/32" with a very thin line*

*The guiding rule for lettering thickness is that as the lettering gets smaller, the line thickness gets thinner. Also notice that drawing titles and subtitles are <u>underlined</u>. The underline should not touch the lettering.*

Lettering from a machine (such as a Kroy® machine) or transfer lettering (such as Letraset®) looks like this and comes in a variety of styles and sizes. Similar type styles can be generated on a computer and printed onto applique film, cut to size, and placed on a drawing.

LEROY LETTERING    LETTERING TEMPLATE

ABCDEFGHIJKLM
NOPQRSTUVWXYZ
1234567890 • abcdefgi
hjklmnopqrstuvwxyz

*Above is a set of letters that the student can trace when in need of lettering greater than 1/4". Outlining the letter is sufficient. Shading the center is too time consuming.*

## LETTERING SPACING

There are three variables related to spacing of lettering. The first variable is the space between letters. The second is the space between words, and the third is the space between lines of lettering. Let's look at each individually.

Leaving proper space between letters is a visual skill. Letters need to have approximately the same amount of **space** between them within a word. The space itself should be just big enough to discern one letter from the next, and the letters should not touch. Certain letter combinations pose problems, such as when two consecutive "A"s or "V"s appear. They create a relatively large area but there is little the drafter can do to avoid the situation. On the other hand when an "A" and a "V" are placed together the drafter has maximum flexibility in creating the appropriate space.

A simple style worth emulating is to place letters so that they almost touch each other. As one gains experience and an "eye" for spacing, this method may be refined into a more personalized method.

The space between words should also be consistent throughout a drawing. The space should range between 1 and 1-1/2 full letter widths. The letter we use for this spacing should be a full one such as an "O".

Spacing between lines of lettering is done according to either of two widely used methods. The first method is to make the space between each line the same dimension as the height of the letters. The second method is to make the space 1/2 of the height of the letters. In other words, if you are using 1/4" lettering, then the space between lines of lettering would be 1/8".

The first method is very fast, but the second method reads a lot better. It is up to the school or employer to establish the standard. For titles and subtitles, which are "free floating" on a drawing, we generally leave room around the line at least as large as the height of the letters. Thus general lettering under a 1/4" title would best be positioned by leaving at least 1/4" between the two lines.

LETTERS ALMOST TOUCH
EACH OTHER
AA  W        AV   AV

*Letters that are made to almost touch each other give a pleasing result. Some letters may leave too much area between them; others leave some flexibility.*

FULL○LETTER
1.5○LETTER

*Leaving the space of one full letter or one full and one narrow letter is just about the right space between words.*

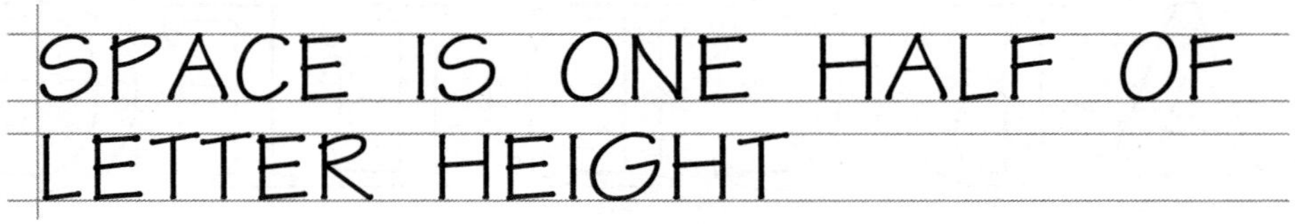

SPACE IS ONE HALF OF
LETTER HEIGHT

*Above is the technique of one half letter height between lines of lettering.*

SPACE IS THE SAME AS
LETTER HEIGHT

*Shown here is the technique of the space between lines of lettering being equal to the height of the lettering.*

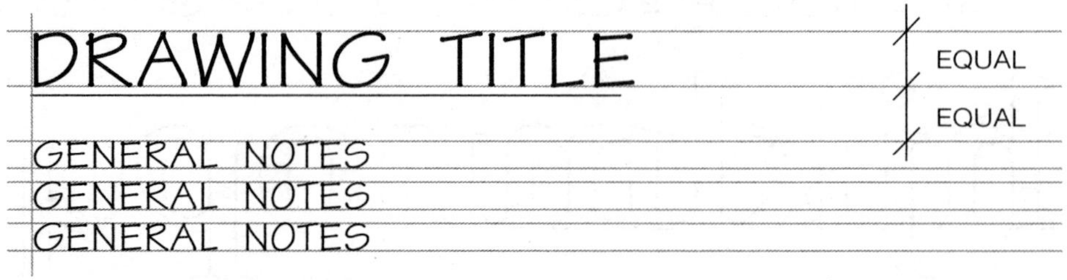

DRAWING TITLE                 ⊢ EQUAL
GENERAL NOTES                 ⊢ EQUAL
GENERAL NOTES
GENERAL NOTES

*When smaller lettering follows larger lettering, especially a drawing title that is underlined, a space equal to the larger lettering should be left between them.*

THIS IS 1/4" HIGH LETTERING, THE SPACE BETWEEN LINES OF LETTERING IS 1/2 X 1/4" = 1/8".

1/4"
1/8"
1/4"

3/16" LETTERING IS SPACED THE SAME WAY, 3/16" X 1/2 = 3/32" SPACING.

3/16"
3/32"
3/16"

1/8" LETTERING FOLLOWS THE SAME RULE, 1/8" X 1/2 = 1/16" SPACING. THIS ONE IS EASY TO SET UP LIKE 1/4" LETTERING BECAUSE OF ITS EVEN DIMENSIONS.

1/8"
1/16"
1/8"

FINALLY 3/32" LETTERING, ALTHOUGH TOUGH TO MEASURE, IS SPACED AT 3/32" X 1/2 = 3/64". AN UPCOMING CHAPTER ON USE OF SCALES WILL SHOW THAT IT IS NOT AS DIFFICULT TO MEASURE AS IT MIGHT SEEM.

3/32"
3/64"
3/32"

THE OTHER METHOD IS TO SPACE LETTERING EQUAL TO LETTER HGT.

1/4" EA.

IT IS POSSIBLE TO DRAW GUIDELINES FASTER WITH THIS METHOD.

3/16" EA.

SOME FEEL THAT THIS SPACING METHOD DOESN'T READ AS SMOOTHLY AS THE OTHER METHOD.

1/8" EA.

BUT WHEN LETTERING GETS SMALL LIKE THIS, THERE IS LESS CHANCE OF ONE LINE RUNNING INTO THE NEXT. THUS THIS TYPE OF SPACING IS GOOD FOR SMALL LETTER SIZES.

3/32" EA.

*The more common technique for spacing lines of lettering is shown above. Many employers prefer it because they feel it looks better and reads better. More lettering can also be fit into a given space.*

*The technique shown above is also used quite a bit. It can be faster to produce guidelines for this technique, and smaller lettering sizes do not tend to run together as they do with the other method.*

# LETTERING PLACEMENT

**Placement** refers to the orientation and location of lettering on a drawing. **Orientation** is the angle at which lettering is placed, and **location** is the precise position where it is placed. Following are guidelines for each type of lettering.

## Titles and Subtitles

Titles and subtitles should be horizontal and read left to right. In CAD terms this is zero degrees. They are commonly aligned with the left edge of the drawing and underlined. The scale is incorporated beneath the underline of the title or subtitle. The scale should be sized about half of the height of the title.

## Text Blocks

There are two types of general notes: *text blocks,* which relate to the whole or major part of a drawing, and notes, which individually identify the components of a drawing. In both cases the orientation can and should be horizontal (zero degrees).

Text blocks can be located wherever there is open space on the drawing. Guidelines on the left (and if needed, on the right) define the limits of the text.

## Material Notes

Notes that define drawing elements should be lettered as close to the objects they are defining as possible. All these individual notes should be aligned with a left hand guideline. The orientation should be either horizontal (0 degrees) or vertical (90 degrees).

Sometimes on plan views, the object that must be noted is oriented at an angle other than 0 or 90 degrees. In this case the lettering should be positioned at the same angle as the object and immediately adjacent to it.

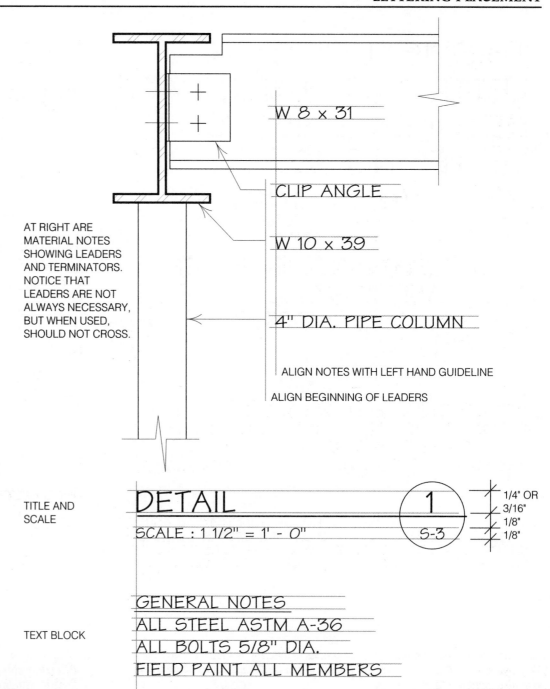

AT RIGHT ARE
MATERIAL NOTES
SHOWING LEADERS
AND TERMINATORS.
NOTICE THAT
LEADERS ARE NOT
ALWAYS NECESSARY,
BUT WHEN USED,
SHOULD NOT CROSS.

W 8 x 31

CLIP ANGLE

W 10 x 39

4" DIA. PIPE COLUMN

ALIGN NOTES WITH LEFT HAND GUIDELINE

ALIGN BEGINNING OF LEADERS

TITLE AND
SCALE

DETAIL                1

SCALE : 1 1/2" = 1' - 0"        S-3

1/4" OR
3/16"
1/8"
1/8"

TEXT BLOCK

GENERAL NOTES
ALL STEEL ASTM A-36
ALL BOLTS 5/8" DIA.
FIELD PAINT ALL MEMBERS

A final but important rule in placing lettering is that it should not touch any object lines. If a note is placed near an object line, the same rule for spacing between lines is applied. In other words, the lettering should be kept away from the line by 1/2 to the full height of the lettering.

The illustrations on this and the facing page show the most common situations encountered when placing lettering:

- titles are set flush left with the drawing
- text is placed within the object it is naming if possible
- if text does not fit within an object, it can be placed above, below, or next to it
- text above or below an object should be separated from it by 1/2 to the full height of the lettering with no leader
- if the text is not too close to the object it is identifying, then a leader line and terminator are needed
- a group of individual notes should be aligned
- text should be placed horizontal if at all possible
- text at an angle should be oriented to the object it names

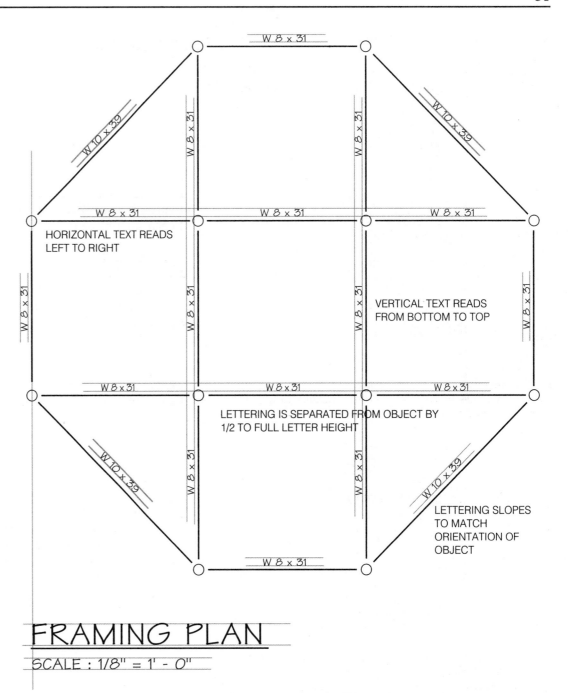

HORIZONTAL TEXT READS LEFT TO RIGHT

VERTICAL TEXT READS FROM BOTTOM TO TOP

LETTERING IS SEPARATED FROM OBJECT BY 1/2 TO FULL LETTER HEIGHT

LETTERING SLOPES TO MATCH ORIENTATION OF OBJECT

FRAMING PLAN

SCALE : 1/8" = 1' - 0"

## Miscellaneous Techniques

There are a couple of special situations that need to be addressed. The following illustrations show them.

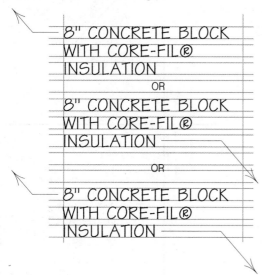

*Leaders and terminators should spring from the beginning of the note or the end of the note. It is also acceptable to spring from both if the object is on both sides of the note.*

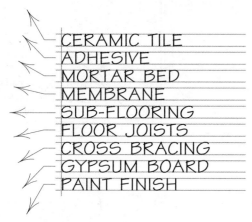

*Multiple leaders and terminators should not cross. By using angles created from basic triangles, leaders can be drawn at 15 degree intervals.*

## LETTERING EXAMPLE

The facing page illustrates a basic drawing incorporating all the variables of proper lettering. Following is an item-by-item checklist a drafter may use to insure proper technique.

Lettering Style
- ❏ only uppercase letters
- ❏ letters vertical
- ❏ letters touch guidelines
- ❏ most letters about as wide as they are high
- ❏ letters such as B, X, and R wider at the bottom
- ❏ each letter made the same each time

Lettering Size and Thickness
- ❏ large title block lettering machine made or hand outlined with a thick line
- ❏ drawing titles at 1/4" with a thick line
- ❏ subtitles at 3/16" with a medium line
- ❏ general notes at 1/8" with a thin line
- ❏ miscellaneous notes at 3/32" with a very thin line

Lettering Spacing
- ❏ equal area between letters
- ❏ space of 1 to 1-1/2 letter width between words
- ❏ space between lines of lettering 1/2 lettering height

Lettering Placement
- ❏ title block contains the project title and address, also design firm's name and address
- ❏ drawing title(s) aligned with left edge of drawing, underlined with scale noted below it at 1/8"
- ❏ subtitle(s) 3/16" high, centered or aligned left, underlined
- ❏ general notes, parts identification: located within field of object or as close as possible, all aligned left together
- ❏ general notes, text block: located in any open area, defined by left and sometimes right guidelines
- ❏ miscellaneous notes: 3/32" high, same as above but for less important information
- ❏ dimensioning: will be covered in depth later, but generally centered

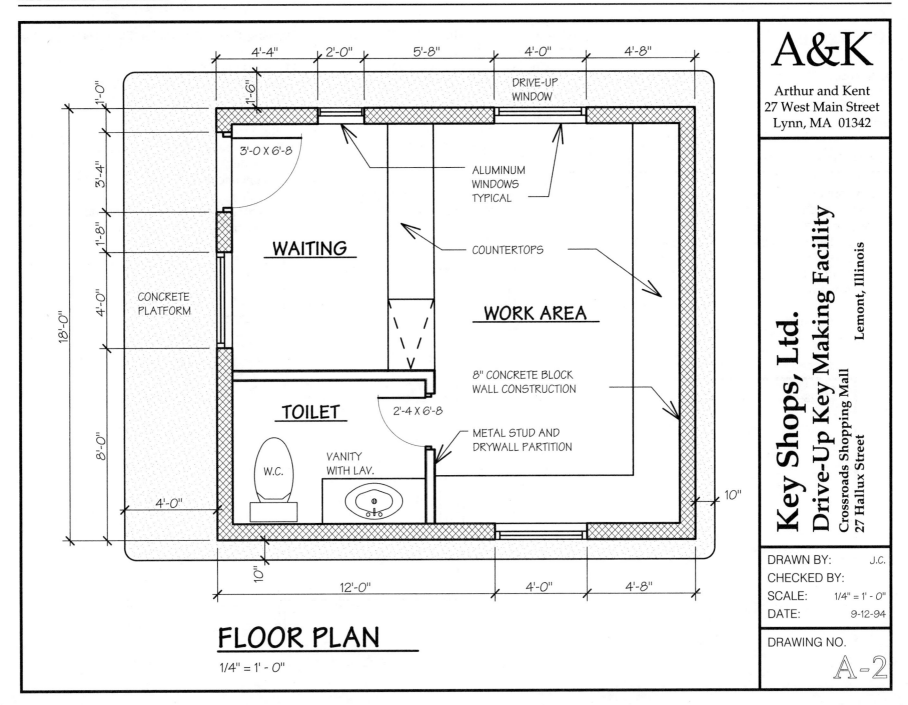

FLOOR PLAN

1/4" = 1' - 0"

# KEY TERMS

uppercase
italic
point sizes
text block

# PRACTICE EXERCISES

1.  Lay out an 8-1/2" x 11" sheet of drafting media and divide it into four areas as shown below. Leave a 1/2" border on all sides. Draw guidelines for lettering at heights 1/4", 3/16", 1/8", and 3/32", one for each area. Practice lettering by copying passages from this text into each of the areas. Letter your name and date above the top border line.

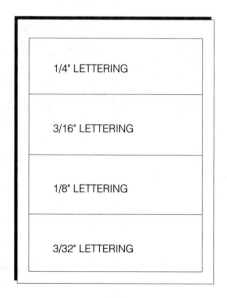

2.  On an 8-1/2" x 11" sheet of media, create a sheet with border and title block. Copy the drawing below but enlarge it by 50%. Include the following lettered information:

Project Title:  Basile Residence Addition
49 Northland Drive
Provo, UT  48201

A/E Firm:  Arch3 Design Group
34 West Springfield Ave.
New London, CT  67498

Drawing Title:  Foundation Plan, 1/4" = 1' - 0"

Materials and Other Notes:
room defined is the "Basement"
12" concrete block (shaded walls)
1' x 2' cont. concrete footing (under wall)
2' x 2' concrete footing (center of basement)
3" dia. steel pipe column (on the 2 x 2 footing)
4" concrete floor slab (the floor within the walls)
"opening to existing basement" (located under the W 8 x 24 lintel)

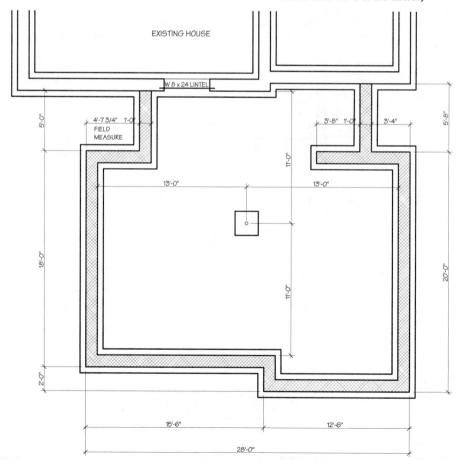

# 5

# SHEET FORMAT AND LAYOUT

## OBJECTIVES

*By the end of this chapter students should be able to:*

- Give standard sheet sizes for the three paper sizing systems used in the drafting industry.
- Identify and create common borders for each of the standard paper sizes.
- Create standard title blocks in their three most common locations on a drawing.
- Identify and practice good sheet layout.

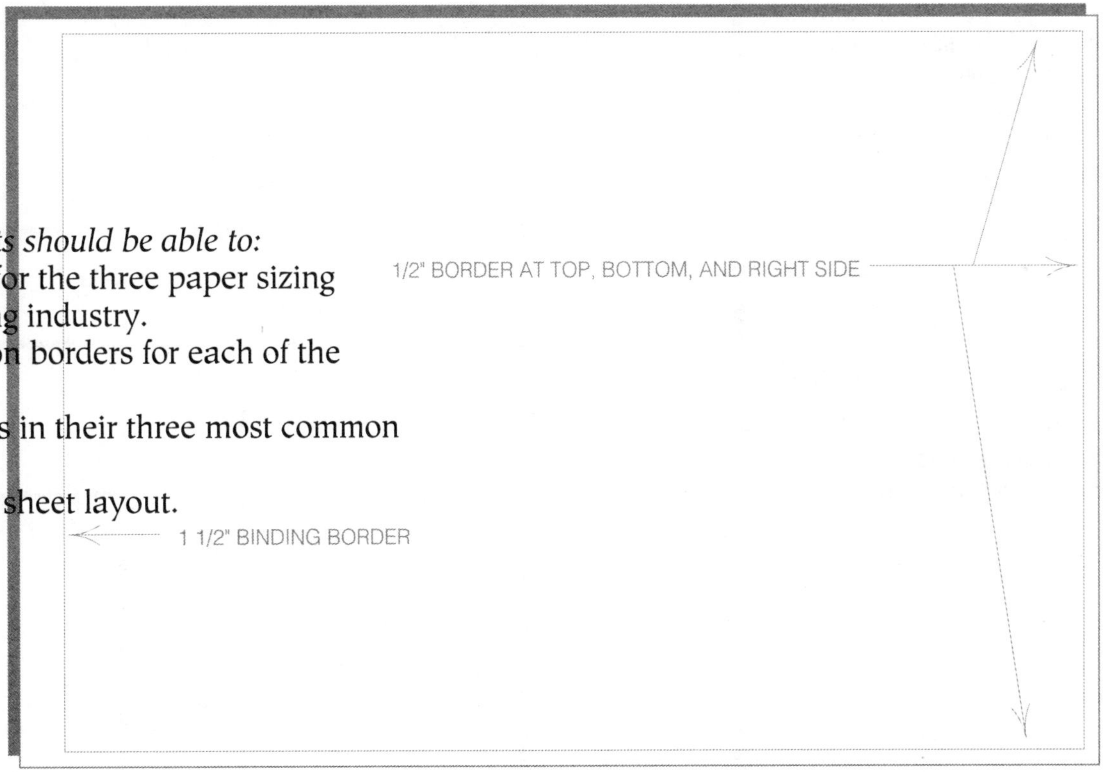

1/2" BORDER AT TOP, BOTTOM, AND RIGHT SIDE

1 1/2" BINDING BORDER

LARGE SHEET LAYOUT

# SHEET SIZES

## ANSI

In the mechanical drafting field there is one recognized sizing system for drawing sheets. This was established and is promoted by the *American National Standards Institute (ANSI)* in an attempt to standardize sheet sizes in the United States.

Some areas of the construction industry have adopted this system, but the majority have not, simply because the sizes just do not accommodate larger objects like buildings and bridges.

Other disciplines have adopted the ANSI system because it is based on a standard 8-1/2" x 11" sheet of paper. Surveyors, for example, prefer it because they can fold their maps down to 8-1/2" x 11" size to be filed with accompanying legal documents.

The first sheet in the ANSI system is a standard 8-1/2" x 11" sheet of paper. That size is doubled four times to create the five standard sheet sizes. Each size is given a reference letter for easy identification as follows:

| LETTER DESIGNATION | SHEET SIZE |
|---|---|
| A | 8-1/2" x 11" |
| B | 11" x 17" |
| C | 17" x 22" |
| D | 22" x 34" |
| E | 34" x 44" |

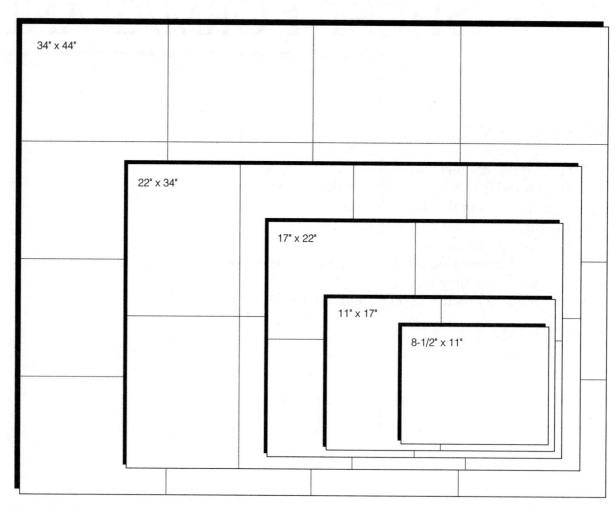

*ANSI Sheet Sizes*

## AEC

The AEC industry has based its system of sheet sizes on a grid and has adopted sizes which accommodate the larger objects they draw.

This system is not an official one, in that it is not recognized by any construction industry organization, but if you were to go to a drafting supplier you would find drafting media and print paper made in sheet sizes based on 1/2 and 1 foot increments.

A popular and widely used size is 24" x 36" (2 ft. by 3 ft.). Occasionally for small projects and shop drawings you will see sheets half that size, 18" x 24". If you were to continue halving that size you come up with the final two small sizes of 12" x 18" and 9" x 12". Even though you can buy these sizes, they are really too small for industry. They are useful in education, however.

For sheets larger than the popular 24" x 36" there are several variations, all multiples of the foot or half foot modular dimension. Thus you can find any combination of 24", 30", 36", 42", and 48". The most popular pre-cut media sizes for the AEC industry are as follows:

> 9" x 12"
> 12" x 18"
> 18" x 24"
> 24" x 36"
> 30" x 42"
> 36" x 48"

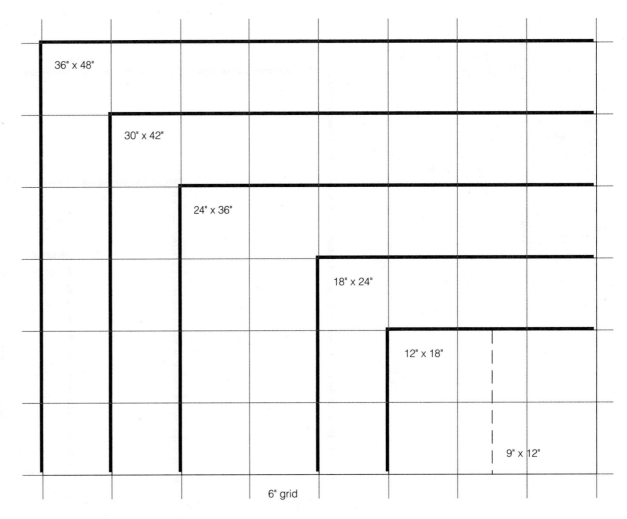

*AEC Sheet Sizes*

# Metric

For most of the civilized world, the system of measurement is the metric system. Measurement of length in the *metric* system is based on the *meter* which is equal to 1,650,763.73 wavelengths of the red-orange radiation in a vacuum of krypton 86.

More practically the meter is just over a yard in length, or 39.37". A *decimeter* is 1/10 of a meter, a *centimeter* is 1/100 of a meter, and a *millimeter* is 1/1000 of a meter.

Sheet sizes for the metric system are broken down like those of the ANSI system. The smallest size is that of a standard metric letter, 210 mm x 297 mm. That size is doubled four times to create the five standard sheet sizes. They are as follows:

| LETTER/NUMBER DESIGNATION | SHEET SIZE |
|---|---|
| A4 | 210 mm x 297 mm |
| A3 | 297 mm x 420 mm |
| A2 | 420 mm x 594 mm |
| A1 | 594 mm x 841 mm |
| A0 | 841 mm x 1189 mm |

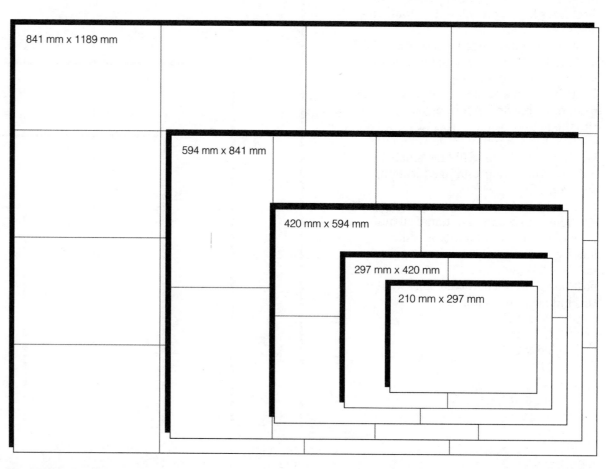

*Metric Sheet Sizes*

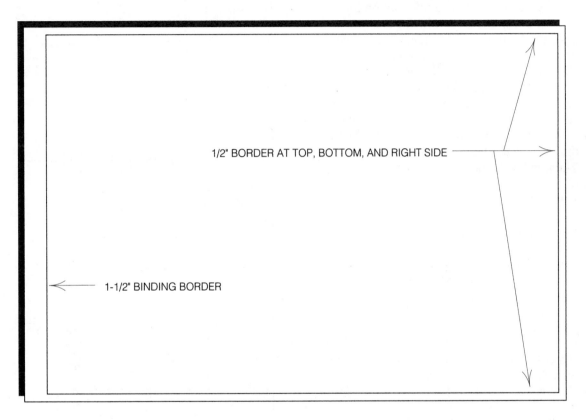

1/2" BORDER AT TOP, BOTTOM, AND RIGHT SIDE

1-1/2" BINDING BORDER

LARGE SHEET LAYOUT

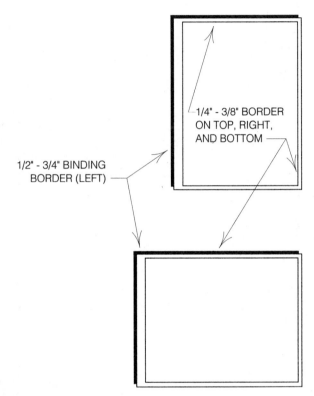

1/4" - 3/8" BORDER ON TOP, RIGHT, AND BOTTOM

1/2" - 3/4" BINDING BORDER (LEFT)

SMALL SHEET LAYOUT

# BORDERS

Most construction drawings should contain a **border**, or margin around the edge of the sheet. This is created using a very thick border line. The border line helps prevent a drafter from drawing right to the edge of a sheet where information could get lost when the drawing is duplicated.

In addition to allowing for duplication, borders also allow space for binding. Most construction drawings are part of a set which must ultimately be bound together. Almost always this is the left edge, so the border line on the left is farther from the edge than the other three sides.

For large sheets the borders are pretty common: 1/2" on three sides and usually 1-1/2" on the binding side, but not less than 1".

For smaller sheets these border dimensions do not leave the drafter with much drawing area. Smaller dimensions for the borders of smaller sheets are therefore acceptable, but usually not less than 1/4" on three sides and 1/2" to 3/4" on the binding side.

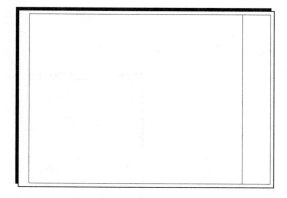

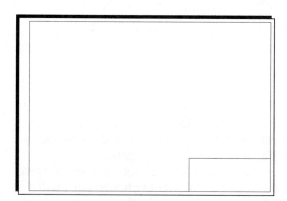

*The three most common locations for title blocks.*

# TITLE BLOCKS

On any project there is some information that is common to every drawing. And like any lengthy document there must be a system of organization to help readers find what they are looking for.

To satisfy these needs, drafters organize this information into defined areas known as *title blocks*. Title blocks in a set of drawings always appear in the same place on a sheet. Usually a company will have its title blocks pre-printed on the drafting media they buy.

The title block should contain information about a project such as:

- project name and address
- company name and address
- sheet title and number
- scale, date, project number
- drafter's name, checker's name
- a place for a licensing stamp
- a list of revisions
- key plan*
- legend*

Location of this information within the title block is not a critical concern except for sheet title and number. This information is meant to help you leaf through the set, so it should be located to the lower right of the title block.

All the other information can be located wherever the designer deems necessary. The guiding thought is to emphasize the more important information while downplaying the least important information.

*\* These items will be discussed later in the text.*

*Sample title blocks showing that although the organization may vary considerably, there is a logical order and a hierarchy of information.*

# LAYOUT

After selecting sheet size, border, and title block, the only question remaining is where to put the drawings. It is a relatively simple task when only one drawing fits on a sheet but a bit more difficult when you have space for three or four. In addition, there is the challenge of arranging the sheets into a set that flows logically. This will be covered later in the text. Following are some generally accepted guidelines for layout.

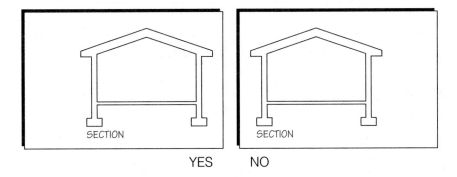

YES         NO

When placing a drawing that does not require the whole sheet, place it towards the right. It is more difficult to look at a drawing the closer you get towards the binding edge, especially with very large sheets.

If there are multiple drawings on a sheet, place the most important one to the right and move towards the left with each succeeding drawing.

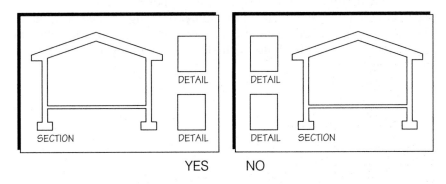

YES         NO

If there is a large drawing and space for small details of the large one, place the small details on the right for ease of viewing.

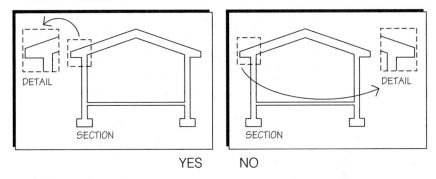

YES         NO

This can contradict the previous guideline but if there is a full drawing on the sheet with details of it, place the details close to where they appear on the large drawing. For example if you are enlarging an area in the upper left hand corner of the main drawing, place that detail adjacent to the main drawing on the left.

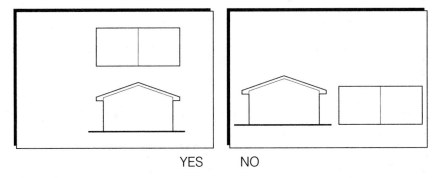

YES         NO

This will be covered further in a later chapter but you should relate different views appropriately. If, for example, you have space on the sheet for a front view of an object and a top view, draw the top view above the front view; as if the front view had been rotated upward. It would be confusing to place a top view below or to the side of the front view.

Likewise if you had a front view and a right side view on a sheet, the correct location would be to place the right side view to the right of the front view. It seems very obvious but it is a common mistake by novice drafters.

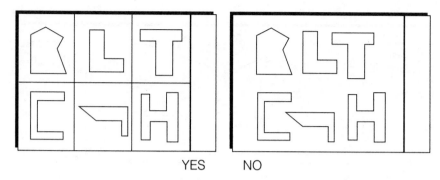

YES       NO

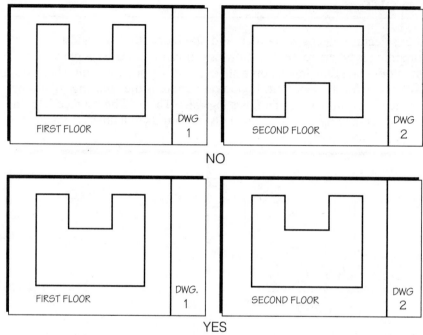

NO

YES

It is also important to separate drawings from each other by enough space so that they are not perceived as one drawing. This can also occur if lettering floats a little too far afield and runs into another drawing.

A grid system may assist in placing drawings on a sheet. Subdividing a sheet with lines is also acceptable.

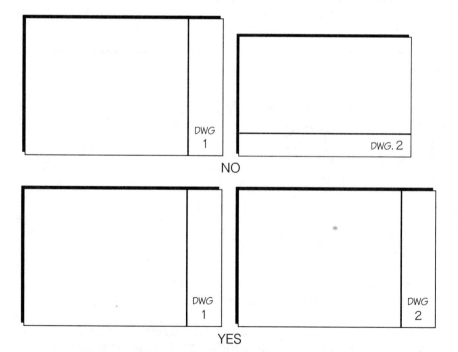

NO

YES

Orient a drawing the same way each time it is drawn. **Orientation** is the direction the object faces in relationship to north. For example, if you draw the first floor plan on one sheet with north facing the top of the page, all other floor plans should face the same way. Or if you enlarge part of an object on the same or different sheet, the orientation should be the same as the mother drawing.

In general, think of drawings as fields of information. Each is a separate entity. It may be related to another drawing on the sheet, but it should be capable of being moved around. This is in fact the situation in CAD drawings where the drafter virtually has the ability to move drawings around at will.

All borders and title blocks should be the same in a set. All sheets in a set should be the same size.

# KEY TERMS

ANSI
metric
meter
decimeter
centimeter
millimeter
border
title block
orientation

# PRACTICE EXERCISES

1. Which title block locations listed below are not common:
   a. upper left hand corner
   b. all across the top
   c. upper right hand corner
   d. all along the right side
   e. lower right hand corner
   f. all across the bottom
   g. lower left hand corner
   h. all along the left side

2. Match the sizes that follow with the appropriate requirement listed below: 1/8", 1/4", 1/2", 3/4", 1", 1-1/2"
   a. binding border for a large sheet
   b. binding border for a small sheet
   c. non-binding border for a large sheet
   d. non-binding border for a small sheet

3. Give the best location or orientation for the following drawings:
   a. left side view when front view is already there
   b. top view when front view is already there
   c. a detail of a large drawing that is already there
   d. a detail of the lower left corner of a drawing that is already there
   e. orientation of a detail

4. Design and draw a title block that you can use for your class projects throughout the term. You may want to set it up for reproduction onto applique film. Design it to fit the sheet size you normally use.

# MEASUREMENT, SCALING, AND DIMENSIONING

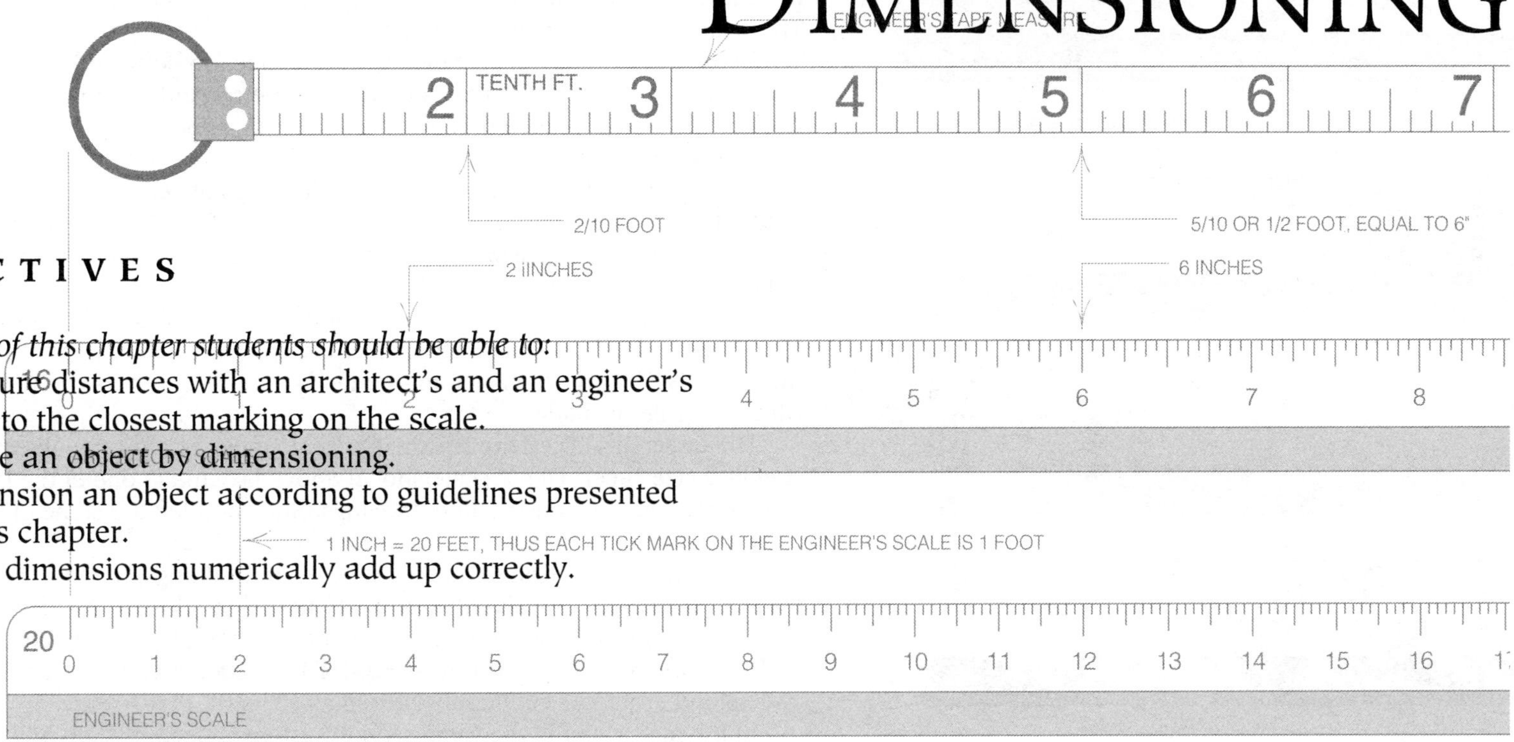

ENGINEER'S TAPE MEASURE

TENTH FT. | 2 | 3 | 4 | 5 | 6 | 7

2/10 FOOT

5/10 OR 1/2 FOOT, EQUAL TO 6"

2 INCHES

6 INCHES

## OBJECTIVES

*By the end of this chapter students should be able to:*

- Measure distances with an architect's and an engineer's scale to the closest marking on the scale.
- Locate an object by dimensioning.
- Dimension an object according to guidelines presented in this chapter.
- Have dimensions numerically add up correctly.

16 | 2 | 3 | 4 | 5 | 6 | 7 | 8

1 INCH = 20 FEET, THUS EACH TICK MARK ON THE ENGINEER'S SCALE IS 1 FOOT

20 | 0 | 1 | 2 | 3 | 4 | 5 | 6 | 7 | 8 | 9 | 10 | 11 | 12 | 13 | 14 | 15 | 16 | 17

ENGINEER'S SCALE

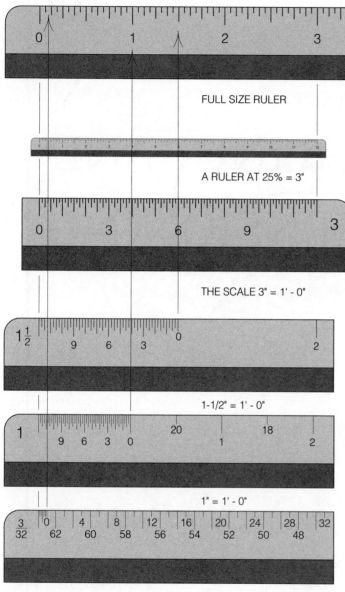

FULL SIZE RULER

A RULER AT 25% = 3"

THE SCALE 3" = 1' - 0"

1-1/2" = 1' - 0"

1" = 1' - 0"

3/32" = 1' - 0"

# SCALE MEASUREMENT

The construction industry deals with very large objects. Beginning with the ground on which a project is built to the individual pieces that comprise the structure, rarely can a drafter draw something at its actual size and have it fit on even the largest sheet of media. For this reason, drawings of construction components must be reduced proportionally. To make them useful they must also be measurable. This is known as drawing them to scale.

In the English system, a standard ruler is one foot long and divided into twelve inches. This page is only 11" wide, so we can either show only part of the ruler or shrink the ruler so that it fits onto the page. Both have been done, above and to the left.

When we need to make construction drawings of very large objects, we need to use a small ruler to do it. This is called a *scale*. The ruler above was reduced to 1/4 or 25% of its actual size. The ruler is now actually 3" long; thus we call it 3" scale or to be more precise, 3" = 1' - 0".

All construction scales are reduced at an even number like this because they would be very difficult to use otherwise. We certainly would not want to reduce a ruler by 37%. It is much easier and perfectly convenient to use increments of which the ruler is made.

The larger an object, the smaller the scale must be to fit the object onto a standard size sheet. The 3" scale shown above is actually one of the largest scales commonly used for construction drawings. Shown at left are some of the other common scales.

Civil Engineers measure objects larger than buildings, namely tracts of land. Even a single piece of residential property usually is close to 200 feet on one side, and subdivisions are measured in thousands of feet. Thus engineers use scales such as 1:20 (read 1 to 20), 1:40, 1:50, 1:100, 1:200, etc., which are more easily understood and more commonly written as 1" = 20', 1" = 40', etc.

In the metric world, the same scaling system is used by both architects and engineers and is based on the meter.

## Architect's Scale

As explained previously, a scale is produced by reducing a standard ruler (one foot) down to a smaller size such as 1/8" or 3/4". An architect's scale is nothing more than a series of lines indicating several of these one foot measurements. The last one foot increment is subdivided into inches. Thus you can measure an object accurately from feet to inches.

This page shows several architectural scales. A vertical line or index goes through all the right reading (top) scales at "0", dividing them into two parts. To the right of the line each scale measures feet, and to the left each measures inches. You can think of these scales as tiny tape measures.

For measuring inches, more tick marks fit into the 1-1/2" scale than the 3/32" scale. It follows that each tick mark on the inch scales represents anywhere from 1/8" to 2" depending on which scale is being used. Larger scales mean drawing with greater detail and accuracy.

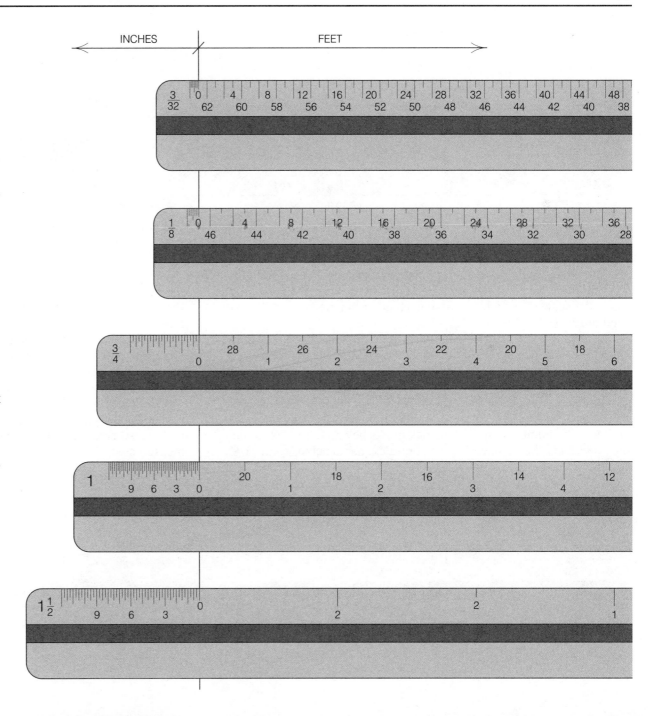

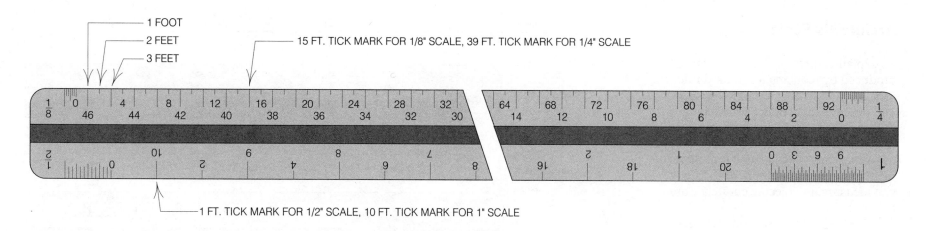

The other problem the drafter must overcome revolves around the fact that there are 11 scales squeezed onto three edges of a standard triangular scale. The scales overlap each other but are compatible with each other.

Look at the 1/8" scale above. Start where the scale is marked 0 and count over four tick marks to the right. You will see the number 4, indicating 4 feet. Now look at the next line to the right. It should be 5 but instead it is 44. Why?

Look closer at all the foot tick marks. They alternate between long and short. The short ones have one set of numbers; the long ones another. The numbers on the long tick marks belong to the 1/4" scale at the other end. The numbers on the short tick marks belong to the 1/8" scale.

This is how two scales overlap; but remember, one is always twice the other. Basically you only need to look at the location of the zero on the scale you are using. It will be aligned with all the other numbers it is related to.

There are many tick marks without numbers on this type of scale, so some interpolation is necessary. For example, on the 1/8" scale, 4 feet is marked and 8 feet is marked, but if you want to measure 5, 6, or 7 feet you must count over to that mark.

The illustration below shows the same distance on two different scales. To measure and draw a line of a given length, simply select the foot measurement on the scale and move down past the 0 to the inch measurement.

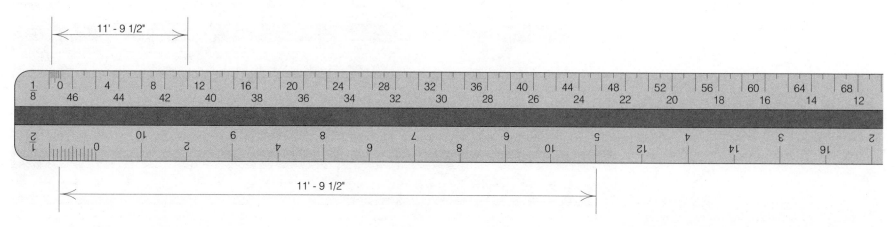

## Engineer's Scale

The engineer's scale is in a practical sense easier to use than the architect's scale, but few students are familiar with its system of measurement. In a way, the engineer's scale is easier to read because there is only one scale per edge—no overlapping of scales.

The engineer's scale does not divide a foot into inches, but rather into tenths or decimals of a foot. This is not to be confused with metric measurement, which is also a system based on tenths. Metric measurement is based on a meter, which is more than three times the length of a foot.

Shown here is a full scale copy of a portion of a tape measure that surveyors and civil engineers use to measure property. The subdivisions

with numbers on them are tenths of a foot, hence the 2 represents 2/10ths of a foot. The tenths are further subdivided 10 times; thus each tick mark is 1/100th of a foot.

Compare the portion of tape measure with the standard ruler illustrated below it. Two inches is not the same as 2/10 foot. **There are no inches on an engineer's tape or scale.** The sooner a student can comprehend that fact, the easier it will be to work with an engineer's scale.

Because engineer's scales are used to measure large distances or dimensions, one cannot measure too accurately with them. With a full size tape measure as shown, it is possible to measure to within 1/100th of an inch. The tick marks on an engineer's scale represent whole feet (or multiples of ten of it) so at best the drafter can only estimate where the exact dimension is.

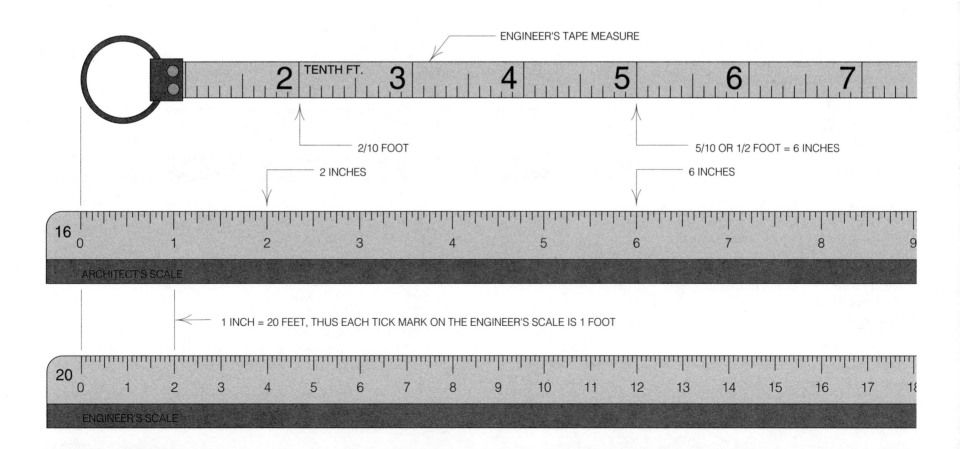

Shown at right is a portion of an engineer's scale. Notice how much cleaner it looks compared to the architect's scale. This is because there is only one scale per edge. Tick marks are numbered every ten units, and there are no tick marks for inches.

An engineer's scale is somewhat flexible. Take the scale 1" = 20 for example. There is nothing to stop the drafter from using that scale as 1' = 20', 1" = 200', or even 1" = 2000'. Each tick mark would then represent 1', 10', and 100' respectively. This is how to draw very large parcels of land on standard size paper.

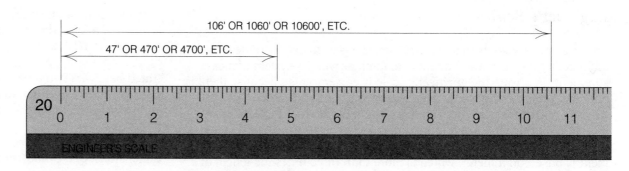

## Metric Scale

Those countries that use the metric system avoid all the confusion associated with two measuring systems. Architects, engineers, and surveyors all use the same system, which is exactly the same as the engineer's except that it is based on a meter, a unit of measure about 39" long.

The meter is subdivided into decimeters (tenths), centimeters (hundreths), and millimeters (thousandths). The corresponding scales of 1:20, 1:25, 1:50, 1:75, etc., are a proportion of actual length to scaled length. The 1:75 scale for example would basically draw an object 1/75th its actual size or 1/750th, etc. Another way to look at it is that .1 meter (10cm) has been reduced in length to represent 75 meters.

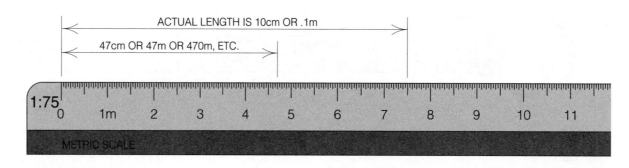

# DIMENSIONING

***Dimensions*** serve two basic purposes:
- they locate an object
- they provide all the measurements necessary to build an object

The first of the two purposes can be accomplished relatively easily. Usually with just two dimensions, an object can be located with respect to a known frame of reference.

The second purpose almost always requires a much more sophisticated system of dimensions to provide all the necessary information without confusing the issue.

As discussed at the beginning of this chapter, it is possible to measure very accurately at full size out in the field. When the drafter draws a reduced representation, however, there are not enough markings on a scale to draw to that precision. Thus the people who read a drafter's drawings cannot be expected to measure sizes from them. To convey the exact measurements of objects, we use dimensions, a systematic way of assigning values for size.

As you may remember from the chapter on lines, dimension lines, extension lines, and terminators are used to create a dimensioning system. Each discipline in the construction industry has its own peculiarities but there are many common principles. Following are the four basic steps of dimensioning.

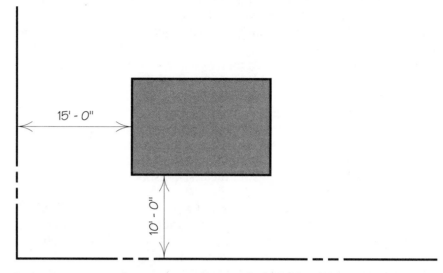

*In the vast majority of situations an object can be located with just two dimensions, one in the X-direction, the other in the Y-direction. In this example a building is located on a piece of land. The land, as defined by its boundaries, is the frame of reference or known location, and the building is positioned in relation to one of its corners.*

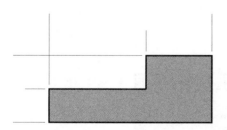 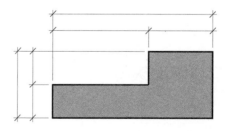 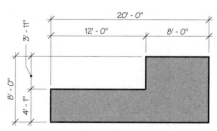 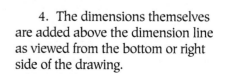

1. Draw extension lines to project from key points of the object. They should not touch the object and should be separate enough to be noticeable, about 1/16".

2. Add dimension lines next. Their locations should follow the guidelines presented on the following pages.

3. Terminators are used to indicate exactly where each dimension begins and ends.

4. The dimensions themselves are added above the dimension line as viewed from the bottom or right side of the drawing.

Like all other aspects of drafting, good dimensioning can be accomplished by drafters putting themselves in the place of the builders. Keep asking yourself what dimensions you would need if you had to build the object yourself.

Look at the simple object on this page and notice the hierarchy of dimensions. If you had to build this from wood you might ask yourself these questions:

- What size piece of wood do I need?
- How big is the notch that must be cut out?
- How big is the cut-out, and where is it located?

The overall dimensions answer the first question for you. (See #1 below.)

The offset dimensions provide information about the notch. And notice that by running those strings the full width and length of the object, you as the builder can measure from either end. (See #2 right.)

With two dimension strings, that question about the cut-out is answered. Notice the efficiency of sizing the cut-out and its distance from the edges by combining a dimension string in each direction. (See #3 below right.)

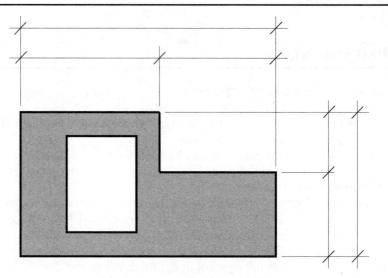

2. Two more lines of dimensions, called **dimension strings**, that size the **offsets** or change of plane of the object are added. These should be placed about 1/2" inside of the overall dimensions.

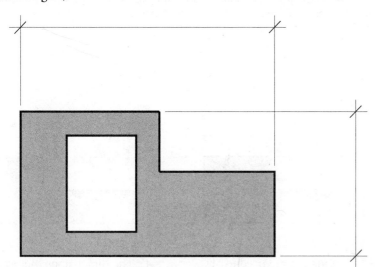

1. The object is sized with two overall dimensions giving its main size. These dimensions need only appear on one side each, but they should be combined with the dimensions explained next.

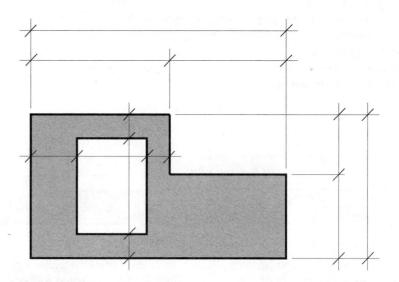

3. Dimension strings for any interior features are added as needed. Usually at least two are needed to define a feature.

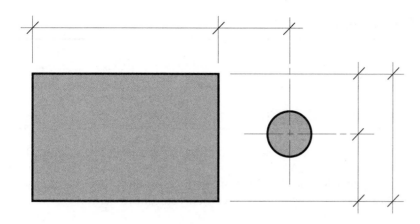

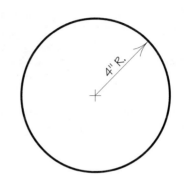

Center lines of objects can and should be used as extension lines. Usually circular or other symmetrical objects are dimensioned to the center line, but as we will see later, other objects are easiest installed when dimensioned to the center.

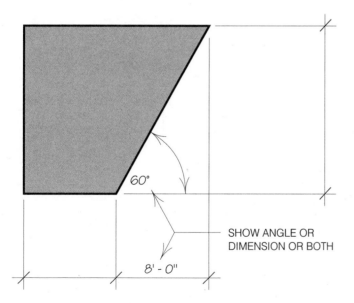

SHOW ANGLE OR DIMENSION OR BOTH

60°

8' - 0"

An angle can be dimensioned with its own or shared extension line and an arcing dimension line. The dimension itself need not follow the arc. It is usually placed by the vertex of the angle.

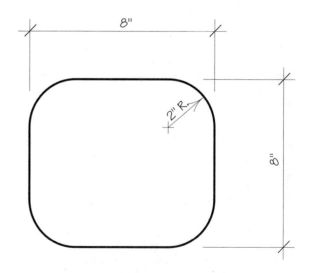

4" R.

8"

2" R.

8"

Radii rather than diameters are usually dimensioned. They are dimensioned by showing the center point with a small "+" and the radius with an arrow. The dimension is shown adjacent to the radius arrow.

Dimensions themselves are placed above their corresponding dimension line. The dimension line is not broken as in mechanical drafting. The orientation of the dimensions follows the rules of lettering; that is, horizontal reads left to right, and vertical reads bottom to top.

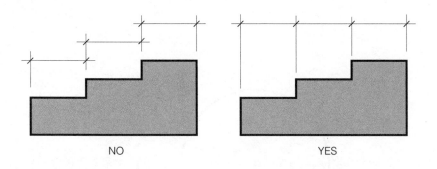

NO                                    YES

When dimensioning, try to combine as many dimensions into strings as possible. This will help the drawing to look less cluttered.

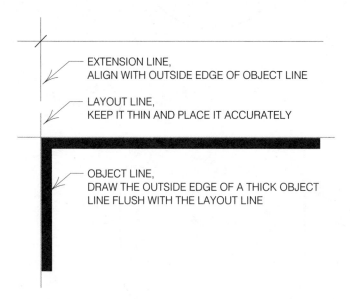

EXTENSION LINE,
ALIGN WITH OUTSIDE EDGE OF OBJECT LINE

LAYOUT LINE,
KEEP IT THIN AND PLACE IT ACCURATELY

OBJECT LINE,
DRAW THE OUTSIDE EDGE OF A THICK OBJECT
LINE FLUSH WITH THE LAYOUT LINE

Most objects' dimensions are taken to their outside edges; therefore the extension line should spring from the outside face of the thick line defining the object, not the center of it. Thus when drawing layout lines, keep them thin for accuracy and then draw the object line up to the edge of them.

# KEY TERMS

scale                                    dimension string
dimensions                               offset

# PRACTICE EXERCISES

1.  Draw the following objects on drafting media using ink or lead. Draw them at any two architectural or engineering scales and to the dimensions you select. Make notes of the dimensions you select and then record those dimensions correctly on your drawing.

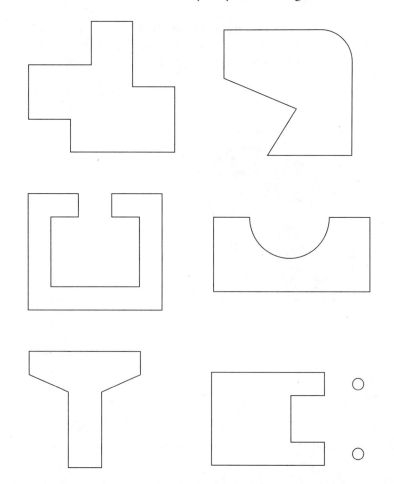

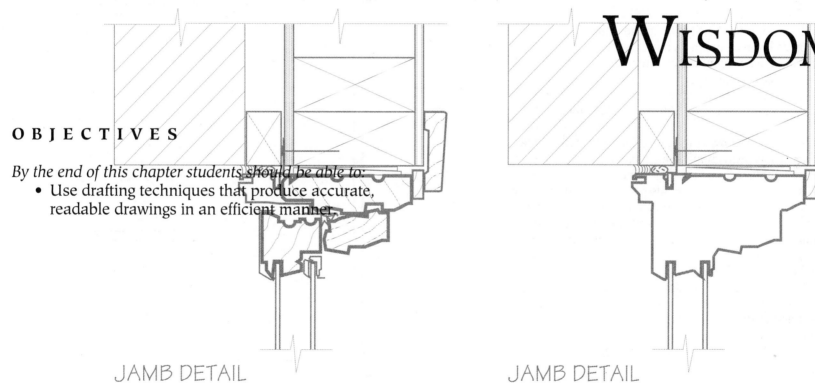

# ACCURACY, EFFICIENCY, AND WISDOM

7

**OBJECTIVES**

*By the end of this chapter students should be able to:*
- Use drafting techniques that produce accurate, readable drawings in an efficient manner.

JAMB DETAIL

JAMB DETAIL

# DRAWING IS A MEANS

*"If you would learn to draw*
  *Hold the instrument often*
    *and hold it with your head."*

William Kirby Lockhard in *Drawing as a Means to Architecture*

While I was a student, I think no other words in any of my texts struck me as this quotation did. It is a reminder that drawing relies on the use of one's brain as much as one's hands.

Drawing is in no way a simple act of recording information on paper with lines. It is a language unto itself with the power to render complex ideas understandable. It is truly a *means* or an instrument to an end, thus the title of the text, *Drawing as a Means to Architecture.*

Drawing is the method by which powerful and creative ideas, that exist only in a few individuals' minds, can be transformed into real three-dimensional monuments for generations to enjoy. A technician or drafter needs to accept this perspective on his or her profession. There is very little room for people who only draw lines.

Drafting requires thinking at all times. It is more accurate to say that a drafter gathers, interprets, and communicates information rather than just records it, because a project is in a state of evolution until the last brick is laid.

Drafting requires accurate and efficient presentation of data. This calls upon a drafter's wit and wisdom. This chapter can only introduce a student to the idea that a drawing instrument is connected to a drafter's mind. The student must take that idea to heart and make it a part of his or her own drafting process.

## TECHNIQUES

With experience, there are many techniques a drafter will discover. Presented here are but a few to stimulate students to begin their own "library" of techniques that will help them create accurate drawings in an efficient manner.

## Consider Detail to Determine Scale

When selecting a scale for a specific drawing, the drafter needs to make a mental adjustment as to what can and cannot be shown. This has to match what needs to be shown on that drawing.

If, for example, a drafter needs to show where a window fits into a wall, it can be done on a fairly small scale floor plan. If the need is to show how it is attached to the wall, then a larger scale will be needed.

Drafters must also provide enough visual information about an object to be able to recognize it. As we saw in the chapter on line work, a thick outline, proper shape, and material indication lines can adequately convey what an object is. As the scale of an object becomes smaller, however, it becomes more difficult to do this.

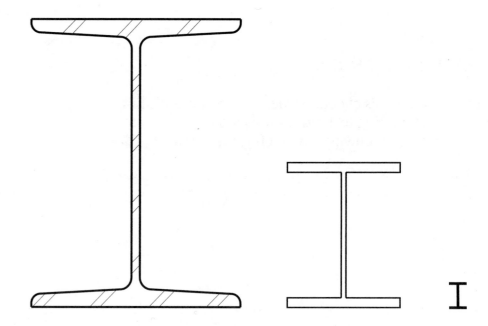

*A section of a steel beam is shown at a large scale. The drawing is accurate down to the radii of the corners and the material indication lines. When the scale is reduced it becomes difficult to draw the radii and the material indication lines, but at least the outline of the shape is clear. Finally at the smallest scale the beam has been reduced to a drawing of three simple black lines. These are all acceptable representations of the object.*

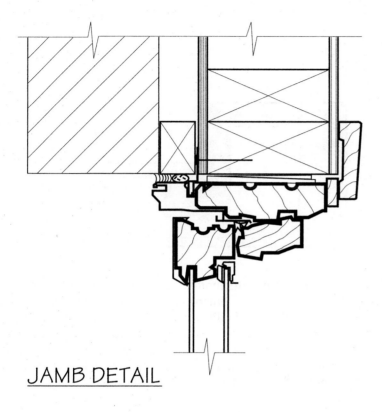

JAMB DETAIL

*A wood window arrives at the site completely assembled and only needs to be installed into the wall. To detail all its parts like this is really a waste of time.*

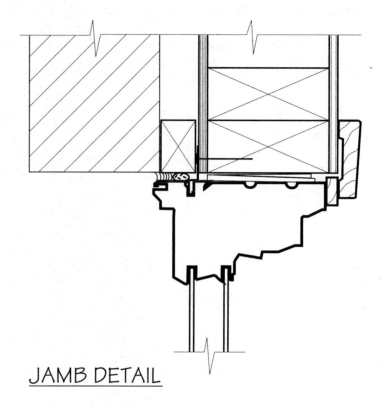

JAMB DETAIL

*An outline of the window such as this is really all the carpenter needs to complete the job. It shows how and where the prefabricated unit fits into the wall.*

## Show Only What Is Necessary

Sometimes there is no need to draw an object in great detail. This occurs when the object being drawn is delivered to the job site already assembled, such as a window, air conditioning unit, or prefabricated fireplace.

The drafter can look in a catalog and find very detailed information on items such as these, but there is no need to detail them on drawings when it is the manufacturer who is assembles them. Again the drafter

needs to examine what the drawing needs to show. In the situation above, the drawing need only show how to attach the window to the wood frame and masonry wall.

On the other hand, if the drafter were drawing a built-in fireplace, surely it would need to be drawn accurately, probably at a larger scale, dimensioned, noted, and otherwise completely described. This must be done because it will be built by tradesmen at the site, not in a factory.

## Use Control to Gain Accuracy

There are several techniques a drafter can use to create an accurate drawing, to make the job easier, and to make the drawing easier to read. A few of the main ones are presented here.

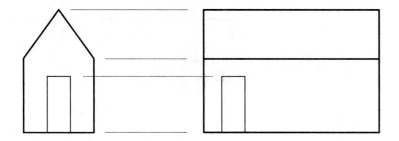

Whenever possible, project rather than measure dimensions from another drawing. Copies of other drawings can get distorted and errors can be made in transferring; when projecting, however, the probability of error is reduced.

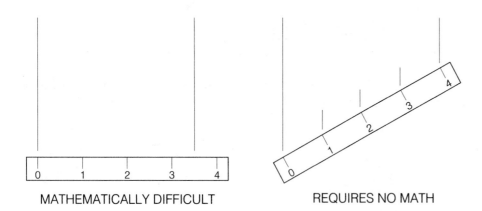

MATHEMATICALLY DIFFICULT            REQUIRES NO MATH

When dividing an odd distance into equal parts, it can be very cumbersome to mathematically calculate the dimension of each part. To divide the distance, simply tilt a scale at whatever angle is needed so that the distance is divided by the required number of spaces.

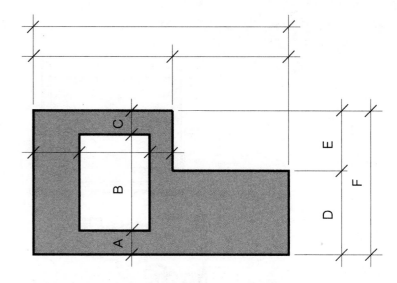

For dimensioning, the drafter must remember that there are usually several strings of dimensions that define the same measurement, but they divide it in different ways. In the example above, there are three vertical dimension strings, and they all measure the same distance. The three strings must add up to the same dimension. Mathematically, this means that A+B+C = D+E = F. It is the drafter's responsibility to check these figures to prevent costly errors at the construction site.

This principle gets to a more basic question of where dimensions come from. The last chapter presented the techniques for measuring and recording dimensions. One could mistakenly infer that a drafter measures a drawing and then records the information. This should never occur.

Drafters must first know what they are drafting. The dimensions of objects either exist or are established before any drawing begins. The drafter then draws the objects to scale and records the actual dimensions. Those who read the drawings later use the written dimensions for accuracy and only use a scale for convenience.

## Approach Drawing as a Process

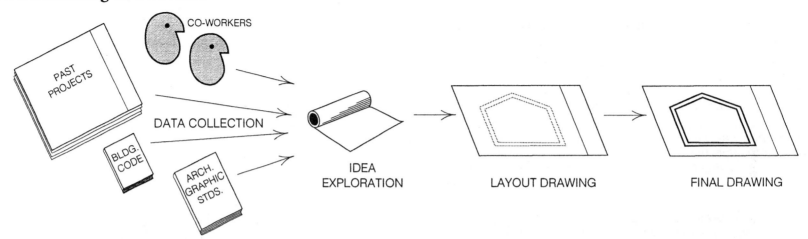

Drawing is a process. Rather than thinking of drawing as making a picture, the student should see it as gathering information, roughing out an idea, laying it out, and then executing it.

The illustration above shows this process. First, the drafter consults other workers, past projects, and reference books. Next, the drafter produces conceptual sketches to focus on how best to present the data. Finally, the drafter produces a layout from which the final drawing can be made.

Let's take the process one step further. The drafter should consider what happens to the final drawing. It is critical that the drafter keep in mind who is going to use the drawing and for what purpose the drawing is intended.

It is the drafter's job to take an idea that exists in peoples' minds and document it so that it communicates the idea to another person. The receiver may be the client, a planning board, another designer, or the contractor. As in the illustration above, the contractor is going to use working drawings to build the project.

Putting oneself in the contractor's shoes while producing the drawings is a worthwhile way to approach the task.

## Follow Generally Accepted Procedures

When a drafter actually creates a drawing, there is a recommended procedure to follow. This procedure illustrates several issues related to the topic of this chapter.

Much of the procedure has to do with cleanliness. A drawing that has been worked on for a few days or weeks can become quite dirty. If it gets too dirty, image quality may suffer to the point that the drawing is unreadable.

The procedure also relates to the hierarchy of information on a drawing. Quite simply the important elements must be drawn before any supporting data are drawn. You cannot draw section lines until the object is defined, and you cannot place text until you know where you have room to put it.

The steps to follow in producing a drawing are as illustrated on this and the facing page. The order of the sequence is not arbitrary. For example, in step 5 the material indication lines are added. This is known as *pochéing*. Even though these lines are very thin, they still should be drawn near the end because many material indications contain a lot of lead work and can produce smudging.

Pochéing is also a time consuming job when done manually. If for some reason the drafter needs to change the position of an object after dimensioning it, no time would have been wasted in pochéing it if the recommended sequence is followed.

The final step of the procedure is to bring all thick and very thick lines up to their appropriate thickness. Thick lines generally represent the edge of an object that has been cut through. When this is the case the work of thickening the outline is called *profiling*. If the drafter were using lead and this step were to be done any earlier, the lead would smudge the drawing during subsequent steps.

Even adding reference symbols earlier would be imprudent. If the sheet on which the detail is being referenced has not been numbered yet, there is a good chance the drafter will forget to go back and add it later. Thus it is better to wait until the set is laid out and add all the reference symbols at one time.

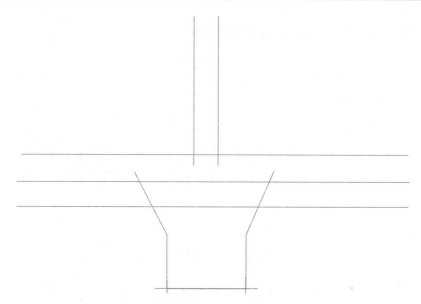

*Step 1: Lay out the entire object to establish the overall size. Only the major parts should be roughed out to avoid duplicating line work later.*

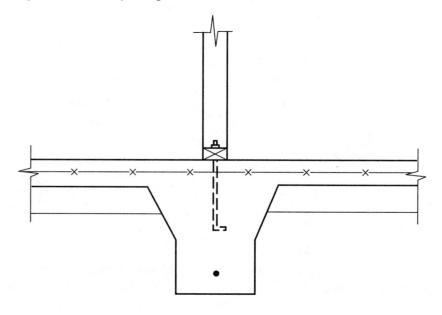

*Step 2: Draw the objects using the correct line thickness up to medium. Do not draw thick and very thick lines yet.*

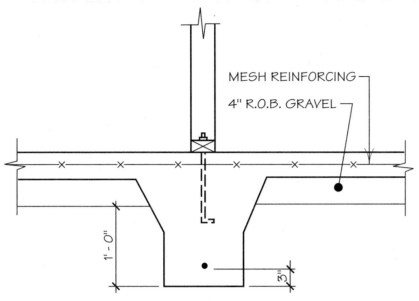

*Step 3: Add dimensions and notes using skills presented in the previous chapters.*

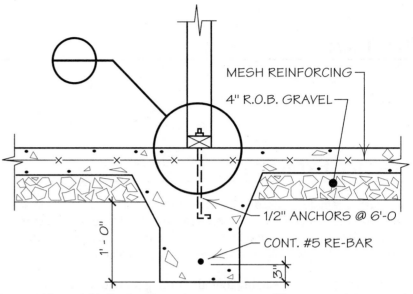

*Step 5: Add material indication lines. It is not necessary to fill an area. Pochéing around the edge or at the beginning and end of a large area is fine.*

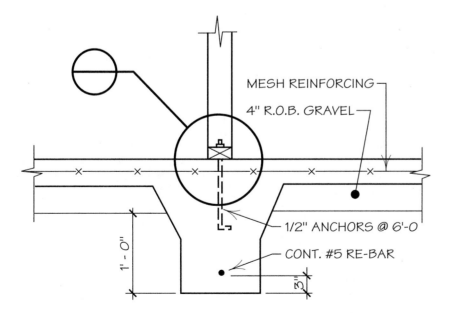

*Step 4: Add reference symbols. This is a simple but important step to aid others who must later read the drawing.*

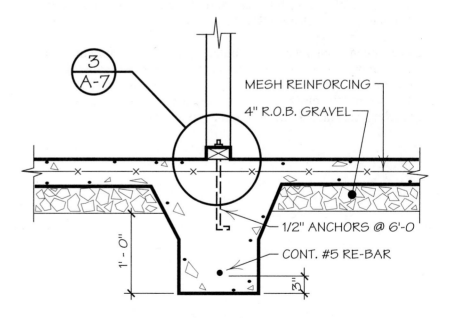

*Step 6: Add thick and very thick lines to finish the drawing. Complete all cross referencing symbols.*

# KEY TERMS

pochéing
profiling

# PRACTICE EXERCISES

1. Measure a material used in construction such as a concrete block,
   section of a window frame, or an item supplied by your instructor.
   Draw the object at three different architectural scales on one sheet of
   8-1/2" x 11" media. Add dimensions to your largest drawing.

2. Measure the door assembly to your drafting room or other nearby
   room. Be sure to measure every aspect of it such as the frame, hinges,
   glass, knob, lock, etc. Draw the door from your measurements at the
   following scales: 1/8", 1/4", 1/2", and 1". Add dimensions to your
   1" scale drawing. Use 8-1/2" x 11" media.

3. Measure an area outside your building such as the entry or a court-
   yard, or other suitable area determined by your instructor. Measure
   the area using an engineer's tape and make two drawings: one at
   1" = 10' and the other at 1" = 50'. Add dimensions to your 1" = 10'
   drawing. Use media size appropriate for the drawings.

# PART II

# CONVENTIONS AND PRELIMINARY DRAWINGS

# MULTI-VIEW DRAWINGS

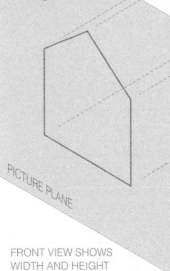

## OBJECTIVES

*By the end of this chapter students should be able to:*
Given a simple object and its dimensions, create accurate
multi-view drawings, including

- plans
- sections
- elevations

TOP VIEW SHOWS WIDTH
AND DEPTH

PICTURE PLANE

PROJECTION LINES

PICTURE PLANE

FRONT VIEW SHOWS
WIDTH AND HEIGHT

PICTURE PLANE

RIGHT VIEW SHOWS
DEPTH AND HEIGHT

# DRAWING CONVENTIONS

One of the challenges a drafter faces is to perceive and depict three-dimensional objects on two-dimensional media. There are several ways to accomplish this task, and we call them drafting or *drawing conventions*.

In general there are four drawing conventions: multi-view, oblique, axonometric, and perspective drawings. The factor that determines which category a drawing fits into is the relationship of the object's surfaces to the viewer, or more specifically the picture plane.

A *picture plane* is an imaginary film placed between the viewer and the object. One can think of the picture plane as the media on which the drawing will be placed or the film in a camera onto which the image is projected. A summary of each drawing type follows.

## Multi-View Drawings

Multi-view drawings actually side-step the problem of showing 3-D objects on 2-D paper by showing only two dimensions at a time. Each face of the object is drawn separately so that proportions and scale can be maintained. Because of this it is impossible to show an entire object with just one view, thus the name multi-view.

*Multi-view drawings* are those in which one face of the object is parallel to the picture plane, and projections from the object are perpendicular to the picture plane.

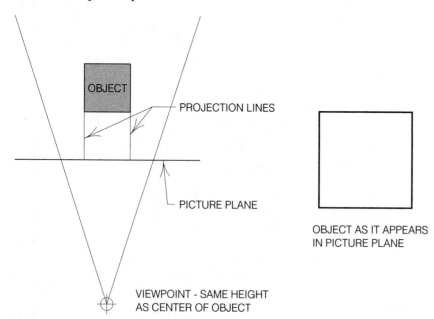

*A front view can easily be generated using multi-view drawings. There is no indication of depth in this drawing however.*

## Oblique Drawings

*Oblique drawings* are those in which one face of the object is parallel to the picture plane, and projections from the object are **oblique** (not perpendicular) to the picture plane. The viewpoint is not only higher than the object, but it is also to the left or right of center of the object. Edge lines along each axis are parallel to each other.

Oblique drawings are the fastest and easiest drawings to produce that show all three dimensions in one view, but they are the least realistic.

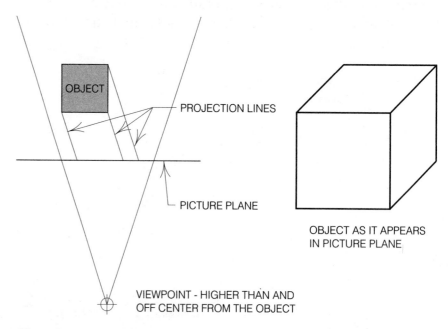

*Oblique drawings provide a simple way of creating the illusion of depth, but they are not very realistic.*

## Axonometric Drawings

*Axonometric drawings* are those in which none of the object's faces are parallel to the picture plane, and projections from the object are perpendicular to the picture plane. The viewpoint is higher than the object, or the object is tipped towards the viewer. Edge lines along each axis are parallel to each other.

Like oblique drawings, axonometric drawings can show all three dimensions or axes in one view. They are a little more involved to produce than oblique drawings, but are closer to what the eye actually sees.

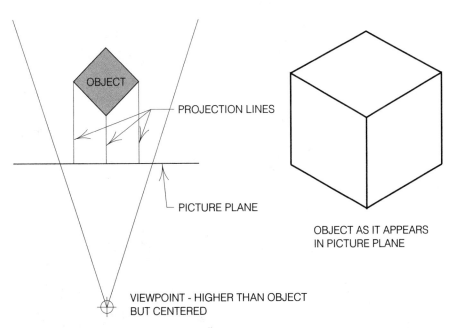

*Axonometric drawings combine reasonable ease of drawing and a view that appears realistic enough for many applications.*

## Perspective Drawings

*Perspective drawings* are those in which a face of the object may or may not be parallel to the picture plane, and projections from the object converge to the viewpoint. These projections are oblique to the picture plane but not parallel to each other as found in oblique views. Edge lines

along the vertical axis may be parallel to each other.

Perspective drawings are the most realistic way of showing all three dimensions in one view. They are equivalent to the view captured by a camera and film. As one might expect, however, they are also the most complex to create.

Oblique, axonometric, and perspective drawings are primarily used to illustrate a project design to the laymen involved in a project. Developers, planning board members, prospective tenants, and investors want to know what the project will look like before they commit to it. A general term for these three conventions is *pictorial views*.

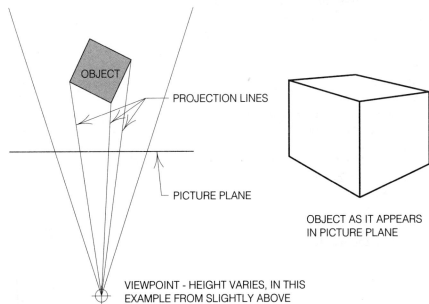

*Perspective drawings give the most realistic view but are also the most complex to draw.*

Since these types of drawings do not show parts of an object in true proportion, they are less useful to a contractor. Contractors need multi-view drawings to build a project. Multi-view drawings are also the most common type produced by AEC drafters. The remainder of this chapter will be devoted to multi-view drawings and the next chapter to the three types of pictorial view drawing.

# ORTHOGRAPHIC PROJECTION

The method of constructing multi-view drawings is called ***orthographic projection***. Orthographic projection assumes that a picture plane is parallel to each face of the object being drawn. Then the image is projected to the picture plane with perpendicular lines.

Therefore, in contrast to the other drawing conventions shown in this chapter, several drawings are needed to illustrate a single object, not just one.

As illustrated on this page, an object is shown in three different views: from the top, from the front, and from the side. It is possible for someone to actually build this object having all three of these views. With only one view, however, it would be impossible. The three views complement each other to fully explain the object.

Multi-view drawings are valuable because they can be drawn proportionally accurate and to scale. When showing only two dimensions per view, it is possible to keep them in perfect proportion. If the length is equal to the width, it will appear that way in the top view no matter how large or small it is drawn.

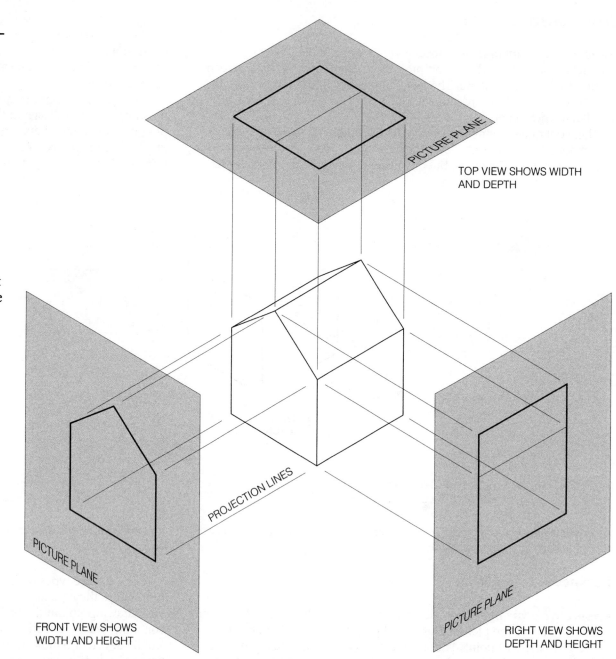

PICTURE PLANE

TOP VIEW SHOWS WIDTH AND DEPTH

PROJECTION LINES

PICTURE PLANE

PICTURE PLANE

FRONT VIEW SHOWS WIDTH AND HEIGHT

RIGHT VIEW SHOWS DEPTH AND HEIGHT

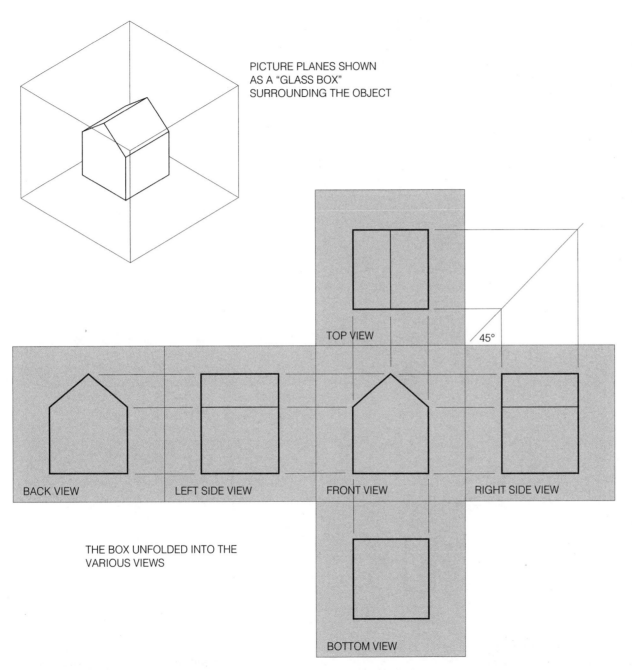

PICTURE PLANES SHOWN
AS A "GLASS BOX"
SURROUNDING THE OBJECT

TOP VIEW

45°

BACK VIEW

LEFT SIDE VIEW

FRONT VIEW

RIGHT SIDE VIEW

BOTTOM VIEW

THE BOX UNFOLDED INTO THE
VARIOUS VIEWS

A common way to understand orthographic projection is to think of the picture planes forming a box around the object. When this "box" is unfolded it creates a series of views.

These views are then in their proper position: the top view is above the front view, the right side view is to the right, the bottom view is below, etc.

Creating multi-view drawings using orthographic projection is fairly easy to do. After the front view is drawn to scale, other views can be constructed with almost no further measurement. All width dimensions can be projected from the front view to produce the top and bottom views, and all height dimensions can be projected to produce the side views. Only the depth needs to be measured on either the top or side view and then transferred to the other by means of a 45° line, which springs from the corner of the front view.

In construction drafting, any view looking down is referred to as a **plan**, rather than a top view. A view of a side is called an **elevation**. There is never any need for a bottom view.

# PLANS

Plans are, in general, views looking down. The type of plan is named after what it is intended to show. The top view of an entire piece of property and its features would be called a **site plan**. The top view of a building would be a view of the roof, or the **roof plan**. The top view of only the building's structure would be the **structural framing plan**. And the top view of the foundation only would be the **foundation plan**.

Since buildings enclose space, a valuable view would be one of the interior looking down on the floor and walls. This is called a **floor plan**. It is perhaps the most important of all views because it shows how the building functions.

To create such a view one would have to cut through the building with a horizontal slice and remove the upper portion of the structure. Such a cut is assumed to occur at about 4 feet above the floor. This would expose most of the interior features.

The floor plan is the most revealing of all drawings and constitutes the largest percentage of drawings in a set. This includes mechanical, electrical, and plumbing plans. And when the building has multiple stories, a plan is needed for each floor, unless the floors are repetitive.

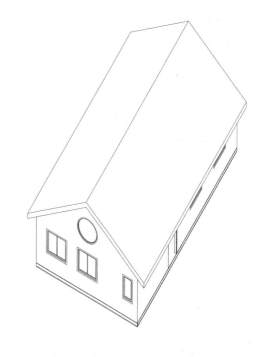

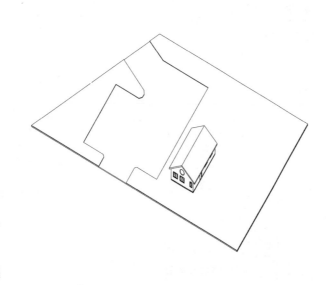

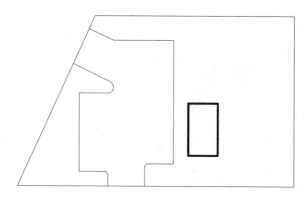

*A top view of a project, including the land it sits on (the site) is called a site plan.*

*A top view of a building produces a roof plan.*

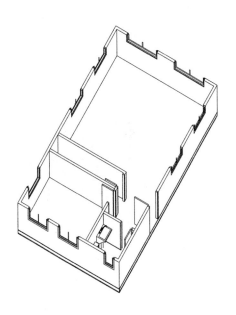

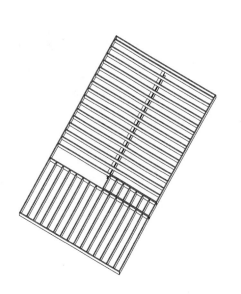

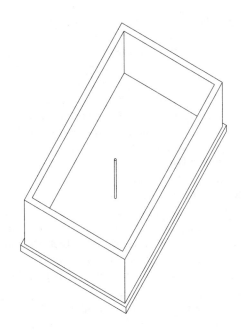

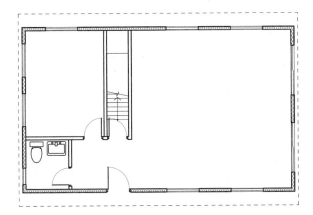

*If a horizontal section is taken through the building, exposing the interior, a floor plan is the result.*

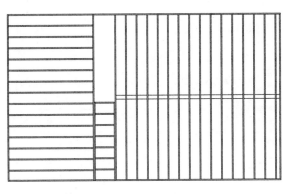

*A view of the structure above the ground is called a framing plan.*

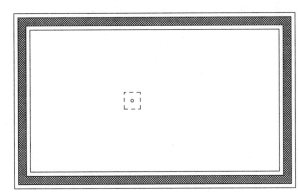

*A view showing the underground support for the building is the foundation plan.*

# SECTIONS

One way to think of a section is by imagining the picture plane going through the object rather than in front of it. A more obvious way of picturing it is to imagine sawing the object in two to reveal its interior. Either way, real structures are seldom seen this way, making this type of drawing a challenge to perceive.

As with plans, there are many types of sections. The name of each section relates to what needs to be shown.

A section through an entire building is called a **building section**. If it is taken down the long axis of a building it is called a **longitudinal section**; and when taken across the short axis it is called a **transverse** or **cross section**. In most cases a cross section of an object provides the most information and is usually easier to comprehend.

While there tends to be a specific number of plan views associated with a design, the number of sections can vary greatly. A simple design might easily be explained with one section. At the other extreme, a design might be so complex that dozens of sections need to be drawn.

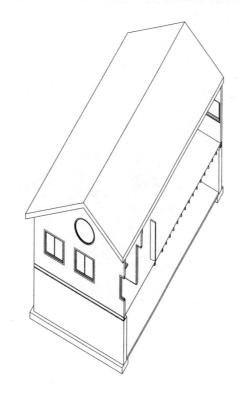

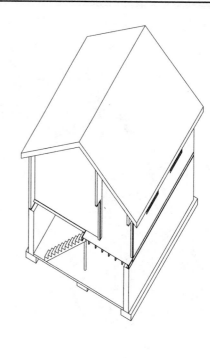

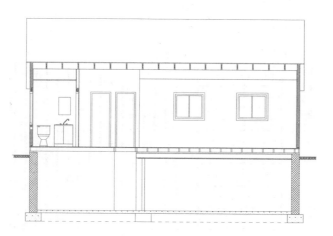

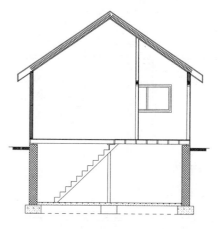

LONGITUDINAL SECTION

CROSS SECTION

Partial sections are also frequently needed when drawing a project. A section through a wall only is called a **wall section**. It can also be seen as an enlarged portion of a building section. A section showing a construction detail is called a **detail section**. It, too, is a further enlargement of a wall or building section.

One can see the range of sections used on a project on this and the facing page. A building section is meant to give an overall understanding. A wall section is to show how major components fit together. A detail section is for showing very precise information.

There are also a group of sections that are combined with elevations for the sake of efficiency.

These include **half sections**, **revolved sections**, and **partial** or **broken-out sections**. Rather than drawing a complete section and a complete elevation, a combination section/elevation can be drawn, thereby reducing drafting time by about a half and still presenting the same information.

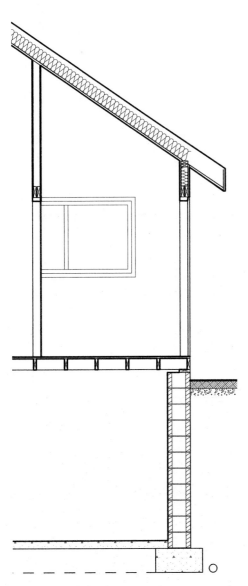

WALL SECTION

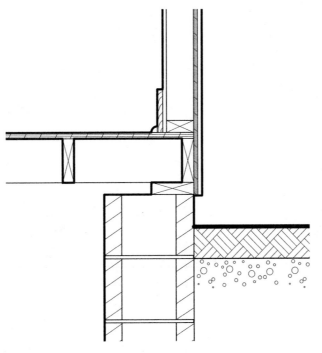

DETAIL SECTION

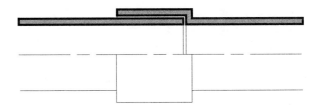

HALF SECTION OF A PIPE CONNECTION

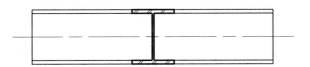

REVOLVED SECTION OF A STEEL BEAM

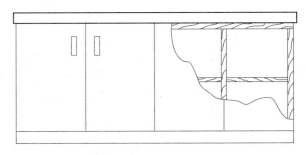

BROKEN-OUT SECTION OF A CABINET

# ELEVATIONS

Elevations are perhaps the easiest views to draw. They may be constructed using the plans and sections already drawn, and they also come close to the views we see in an actual structure. When only the outside is shown, they are referred to as *exterior elevations*.

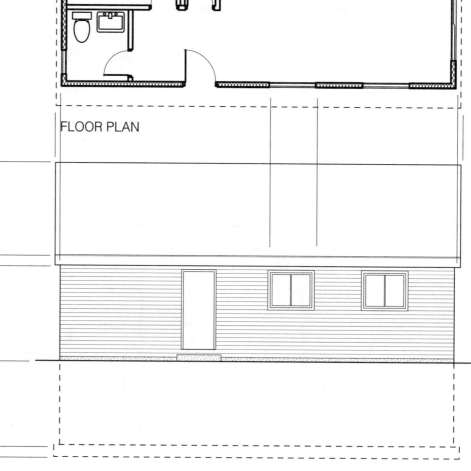

FLOOR PLAN

BUILDING SECTION                                                                    FRONT ELEVATION

It is quite common to show interior surfaces to describe a project. These are known as ***interior elevations***. Complete rooms may be shown in this fashion by showing all four walls.

Each elevation is given a number or letter designation to identify it. This is usually accompanied by a key plan to indicate where the elevation is located.

Both exterior and interior elevations are named and noted somewhat differently depending on the discipline and type of project. Part III of the text will elaborate on elevations as well as all the other views.

For all multi-view drawings it is easiest to draw one view by measurement and then project lines from this view to construct the other views. Most frequently the drafter begins with a plan view, creates vertical sections, and completes the outside views or elevations. This is so because it is the natural sequence of thought when trying to understand a three dimensional object. It is also the sequence that occurs when initially designing the structure.

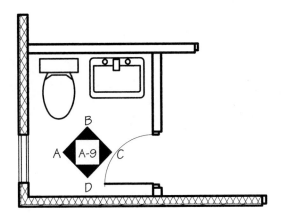

KEY PLAN

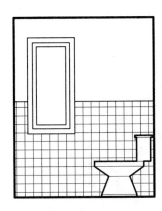
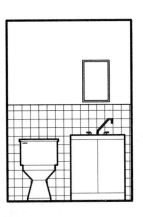
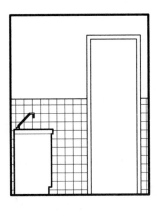
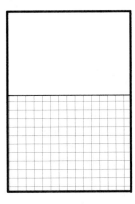

INTERIOR ELEVATION  A          B          C          D

# KEY TERMS

drawing convention
picture plane
multi-view drawing
oblique drawing
axonometric drawing
perspective drawing
pictorial views
orthographic projection
plan
  site plan
  roof plan
  structural framing plan
  foundation plan
  floor plan
section
  building section
  longitudinal section
  transverse section
  cross section
  wall section
  detail section
  half section
  revolved section
  partial section
  broken-out section
elevation
  exterior elevation
  interior elevation

# PRACTICE EXERCISE

Create multi-view drawings for the
following objects (roof plan, floor plan,
all four elevations, one or two sections). The numbers indicate proportions. Select a scale to fit the size
drafting media you normally use.

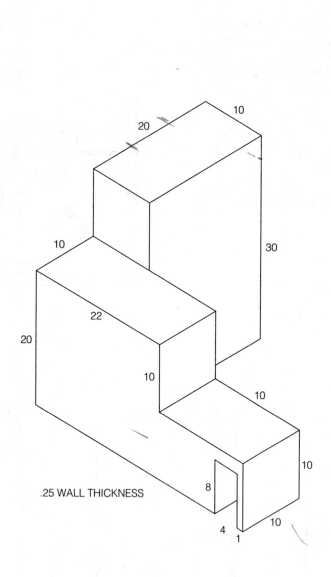

.25 WALL THICKNESS

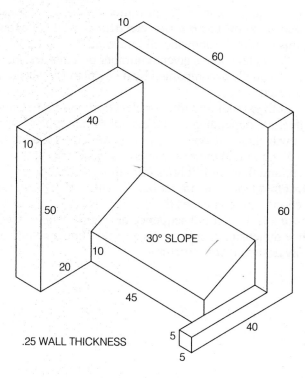

.25 WALL THICKNESS

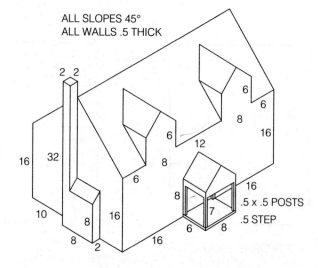

ALL SLOPES 45°
ALL WALLS .5 THICK

.5 x .5 POSTS
.5 STEP

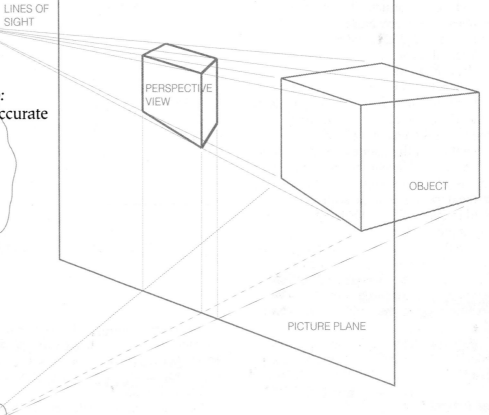

# 9

# PICTORIAL VIEWS

## OBJECTIVES

*By the end of this chapter students should be able to:*
Given a simple object and its dimensions, create accurate
- oblique drawings
- axonometric drawings
- one- and two-point perspectives

LINES OF
SIGHT

PERSPECTIVE
VIEW

VIEWER

OBJECT

PICTURE PLANE

# PARALINE DRAWINGS

Oblique and axonometric drawings are fast and easy ways of producing somewhat realistic 3-D views that are based on multi-view drawings. They are often referred to as *paraline drawings* because edges along each axis are drawn parallel to each other.

## Oblique Drawings

Oblique drawings are drawings of an elevation, or less commonly, a section to scale. The top and one side are constructed by extending parallel lines back at any angle. Standard triangle angles of 30°, 45°, and 60° are the easiest. These lines are not perpendicular to the picture plane, but rather oblique.

All vertical lines on the object are drawn as vertical lines, and all horizontal lines parallel to each other on the object are drawn parallel. Thus they are easy drawings to produce.

The problem with oblique drawings is that while the width and height are drawn to scale, the depth is difficult to proportion. To look correct, the depth needs to be scaled down compared to the width and height. At best we can only approximate it.

The top three drawings on this page represent ways of determining depth by actually measuring with a scale. The *cavalier oblique* drawing uses full scale depth, which is least realistic but useful for drawing very shallow objects.

The *general oblique* drawing has pleasing proportions by measuring back 3/4 the scale of the elevation. If the elevation was drawn at 1/4" scale, for example, the depth would be created using 3/16" scale.

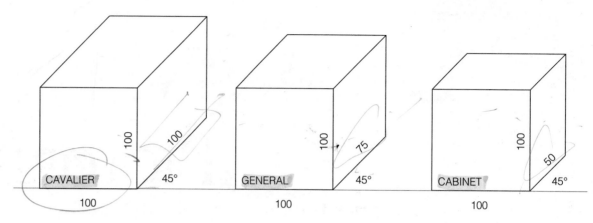

ELEVATION OBLIQUES - MEASURED METHOD

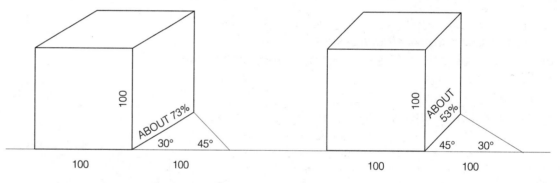

ELEVATION OBLIQUES - PROJECTION METHOD

The *cabinet oblique* drawing perhaps looks most correct and is created with the depth drawn at 1/2 the scale of the elevation. A 1/4" scale elevation would therefore be projected back at 1/8" scale.

Obliques can be created using what is called the projection method, illustrated directly above. The drafter scales over to the right, the same dimension as the object width. The back edge is determined by the meeting of 30° and 45° projections as shown. This produces depths of about 73% and 53%, which are both fairly realistic.

## Axonometric Drawings

Axonometric drawings differ from oblique drawings in that none of an object's faces is parallel to the picture plane. Like oblique drawings, however, some axonometric drawings can be created from an already completed multi-view drawing, such as a plan.

Basically there are three types of axonometric drawings: *isometric*, *dimetric*, and *trimetric*. Their names come from the number of axes that are measured at the same scale. In each case only the top (as in the examples at right) or bottom corner of the object touches the picture plane.

Isometrics are by far the easiest to produce because all three axes make equal angles with the picture plane and share the same scale. "Isometric" means equal measure.

Dimetrics have only two axes at the same scale and same angle to the picture plane. The vertical dimension is foreshortened as shown at right. When the axes of the base are set at right angles, the drawing in essence begins with a plan view.

Dimetrics are constructed much the same as elevation obliques, which is why they are commonly referred to as plan oblique drawings. Rather than projecting back from an elevation, the drafter projects upwards from a plan.

Trimetrics have all three axes at different angles to the picture plane, and thus each is drawn at a different scale. In practice, however, it is common to measure the two horizontal scales the same, i.e., 100%. The axis that is 60° to the picture plan is really slightly less than 100% but is assumed to be so for the sake of simplicity.

With this concession, trimetrics are essentially the same as dimetric drawings because the base of the object is a plan. The only difference is that the base is rotated at a 30-60° angle rather than a 45-45° angle. Other angles are possible for both dimetric and trimetric drawings but they are more difficult to construct and are therefore seldom used.

For both dimetric and trimetric drawings the heights are determined using the same methods shown on the previous page, either measured or projected. Vertical lines are drawn vertically for all axonometrics. Horizontal lines parallel in the object remain so in the drawings.

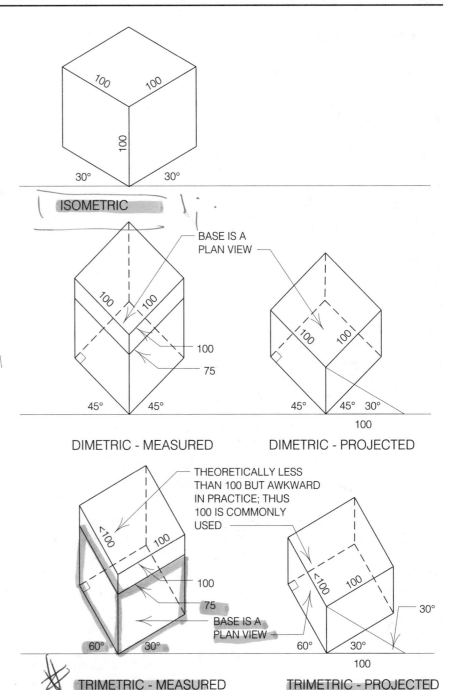

# PERSPECTIVE DRAWINGS

Perspective drawings most closely resemble what the eye actually sees. They show all three axes of an object in one view, and the edges of the object are not parallel to each other.

In perspective drawings, parallel edges on the object are drawn so as to converge to points in the distance called *vanishing points*. Perspectives can have one, two, or three vanishing points, depending on the position of the viewer (*station point*) and the object.

The illustration on this page shows a viewer, an object, and the picture plane. Each edge of the object is seen by the viewer along a line of sight. These lines of sight converge at the viewer's eye and intersect the picture plane at precise locations. It is these points of intersection that create a perspective drawing.

If the viewer is standing far enough back from the object to see all of it, the object is said to fall within the viewer's field or *cone of vision*. This is a relative factor because it varies from person to person, but a cone of vision between 30° and 60° is normal.

Where the viewer stands and how the object sits in relation to the picture plane determines what the perspective will look like.

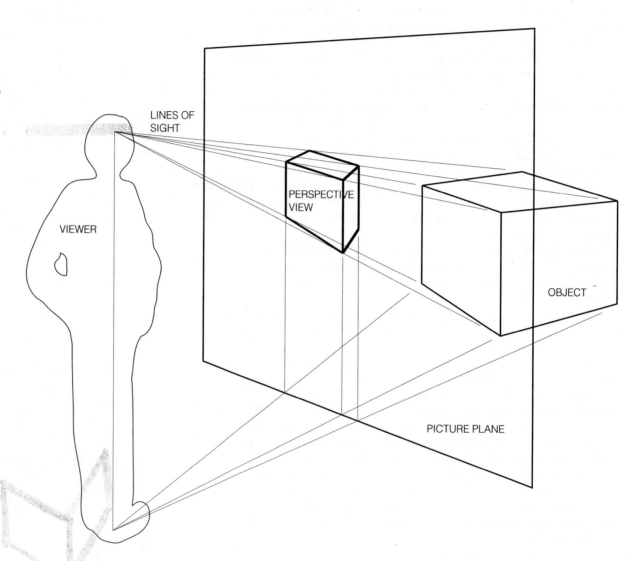

*The concept of a perspective may be new to the student, but comparing it to photography makes a lot of sense. In each case the viewer sees an object that is captured on a plane. The picture plane can be thought of as the film in the camera. The image changes depending on the distance and angle of the viewer to the object.*

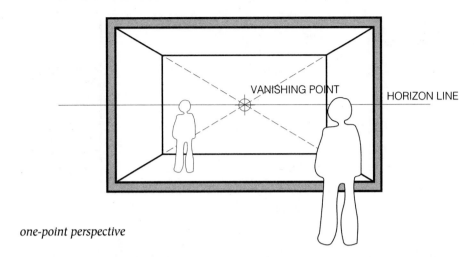

*one-point perspective*

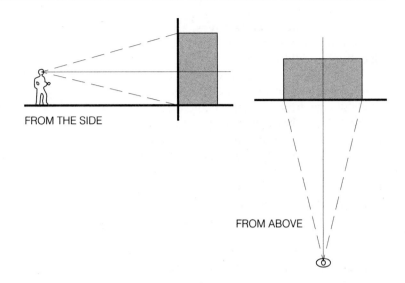

In **one-point perspectives**, all lines perpendicular to the picture plane converge in the distance to a single point. This is the case when the face of the object is parallel to the picture plane and the line of sight is perpendicular to it. The viewer need not be centered on the object. All vertical lines appear and are drawn vertically. This type of perspective occurs naturally when viewing interior spaces and the line of sight is perpendicular to one wall.

*Orientation of viewer (station point), picture plane, and object to create a one-point perspective.*

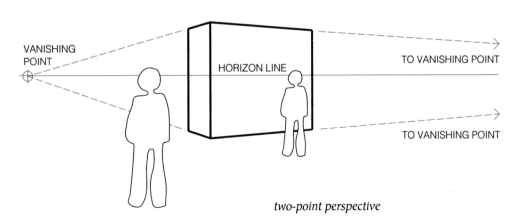

*two-point perspective*

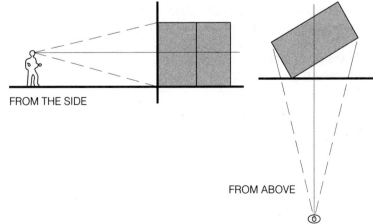

In **two-point perspectives**, vertical lines are also drawn vertically but now horizontal lines converge to two vanishing points. This is the case when showing exterior forms and no face of the object is parallel to the picture plane.

*Orientation of viewer (station point), picture plane, and object to create a two-point perspective.*

In ***three-point perspectives***, all conditions are met for a two-point perspective, but in addition the viewer must look up or down to see all of the object. Thus whenever the viewer needs to change the angle of view from horizontal, and no face of the object is parallel to the picture plane, a three-point perspective will result.

If the viewer looks from above or below, the view also changes, creating perspectives known as "aerial," "bird's eye," and "worm's eye" views for obvious reasons. The possibilities are endless but essentially the concept is the same—a perspective is created by the intersection of lines of sight with the picture plane.

Unlike paraline drawings, the lines of a perspective drawing converge to a person's eyes and therefore create converging lines on the drawing.

Perspectives do not suffer from as many distortions as oblique and axonometric drawings do. They are fairly accurate two-dimensional pictorial views of 3-D objects. In fact, by adding people, shades, shadows, and textures to the drawing, a perspective can rival a photograph.

The two most widely used perspectives are the one-point and two-point. The following pages present step-by-step procedures for creating each.

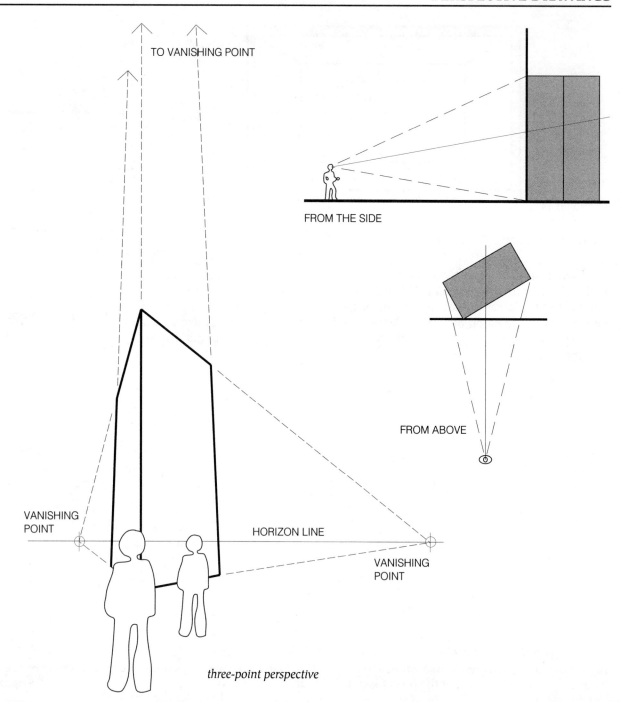

TO VANISHING POINT

FROM THE SIDE

FROM ABOVE

VANISHING POINT

HORIZON LINE

VANISHING POINT

*three-point perspective*

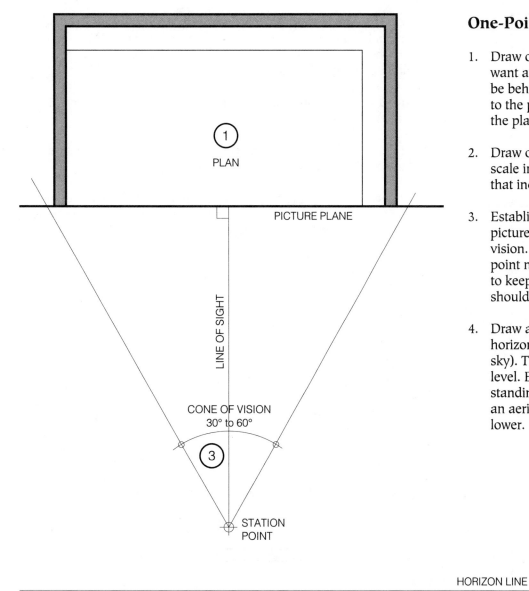

## One-Point Perspective

1. Draw or place a plan of the object at the top of the sheet. If you want an interior view, leave off the part of the plan that would be behind the viewer. Set the plan such that its face is parallel to the picture plane. Set the picture plane at the front edge of the plan.

2. Draw or place an elevation or section of the object at the same scale in the lower right hand corner of the sheet. Choose one that includes important heights.

3. Establish the station point. Set it far enough away from the picture plane so that the plan fits within a 30° to 60° cone of vision. Many consider 45° to be the maximum. The station point need not be centered, but if you move it off center be sure to keep the plan within the selected angle. The line of sight should be perpendicular to the picture plane.

4. Draw a horizontal line through the section. This will be the horizon line (the line in the distance where the earth meets the sky). The most common place to set the horizon line is at eye level. Eye level is simply the distance from the surface you are standing on to your eye, which is 5 to 6 feet for most people. For an aerial view, set it higher; if you are below the object, set it lower.

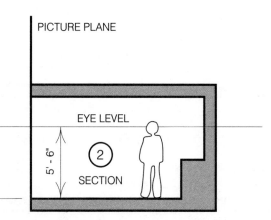

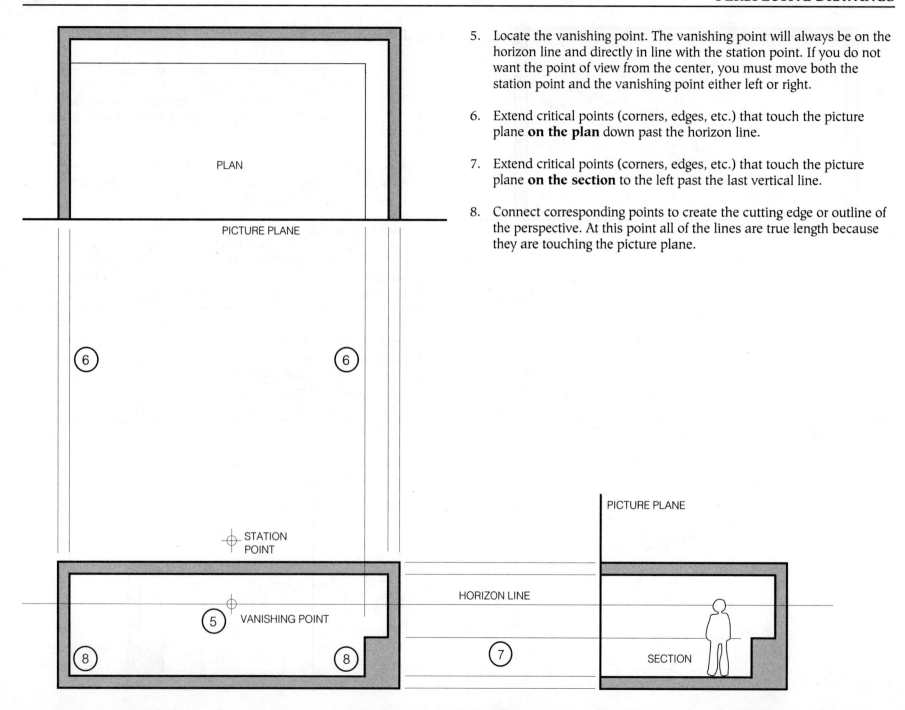

5. Locate the vanishing point. The vanishing point will always be on the horizon line and directly in line with the station point. If you do not want the point of view from the center, you must move both the station point and the vanishing point either left or right.

6. Extend critical points (corners, edges, etc.) that touch the picture plane **on the plan** down past the horizon line.

7. Extend critical points (corners, edges, etc.) that touch the picture plane **on the section** to the left past the last vertical line.

8. Connect corresponding points to create the cutting edge or outline of the perspective. At this point all of the lines are true length because they are touching the picture plane.

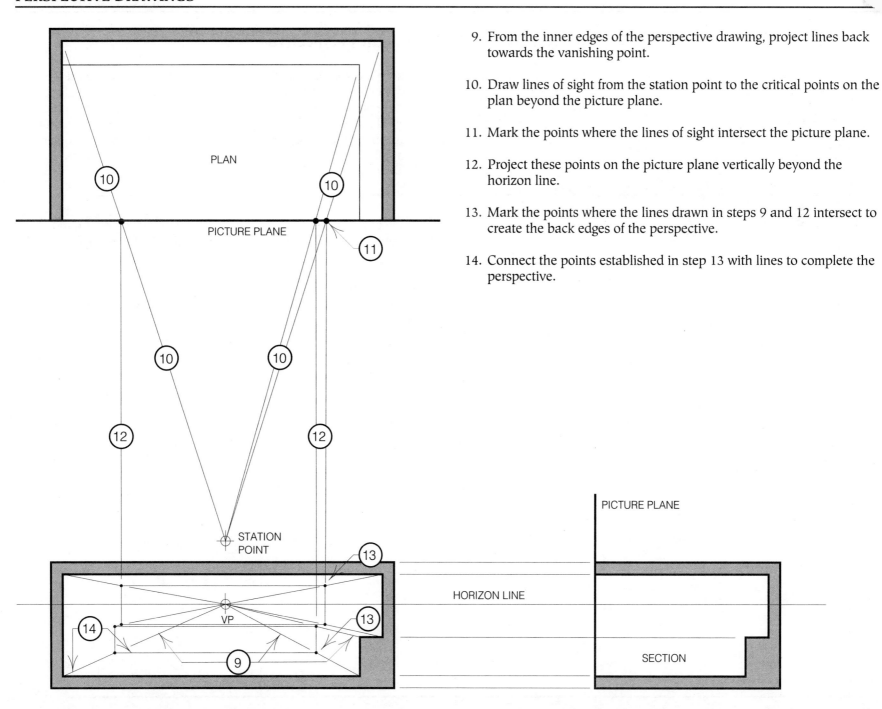

9. From the inner edges of the perspective drawing, project lines back towards the vanishing point.

10. Draw lines of sight from the station point to the critical points on the plan beyond the picture plane.

11. Mark the points where the lines of sight intersect the picture plane.

12. Project these points on the picture plane vertically beyond the horizon line.

13. Mark the points where the lines drawn in steps 9 and 12 intersect to create the back edges of the perspective.

14. Connect the points established in step 13 with lines to complete the perspective.

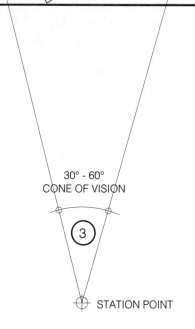

## Two-Point Perspective

1. Draw a picture plane and the object at the top of the sheet. It is common to set the longer dimension of the plan at a 30° angle to the picture plane; however, any angle will work. The plan should be behind the picture plane to avoid distortion in the perspective but at least one point of the object should touch it. An "L" shaped object, for example, could have two corners touching the picture plane.

2. Draw or place an elevation of the object in the lower right hand corner of the sheet. Choose an elevation that includes important heights.

3. Establish the position of the station point. Set it far enough away from the picture plane so that the plan fits within a 30° to 60° cone of vision. The station point need not be centered, but if you move it off center be sure to keep the plan within the selected angle.

4. Draw a line beginning at the station point, parallel to the right hand face of the object, and extend it to the picture plane up and to the right. This line is 30° in this example. Repeat the process for the left hand side at its angle of 60°.

5. Draw the horizon line in relation to the elevation. The most common place to set the horizon line is at eye level. For this example, however, we will use a bird's eye view and locate it higher.

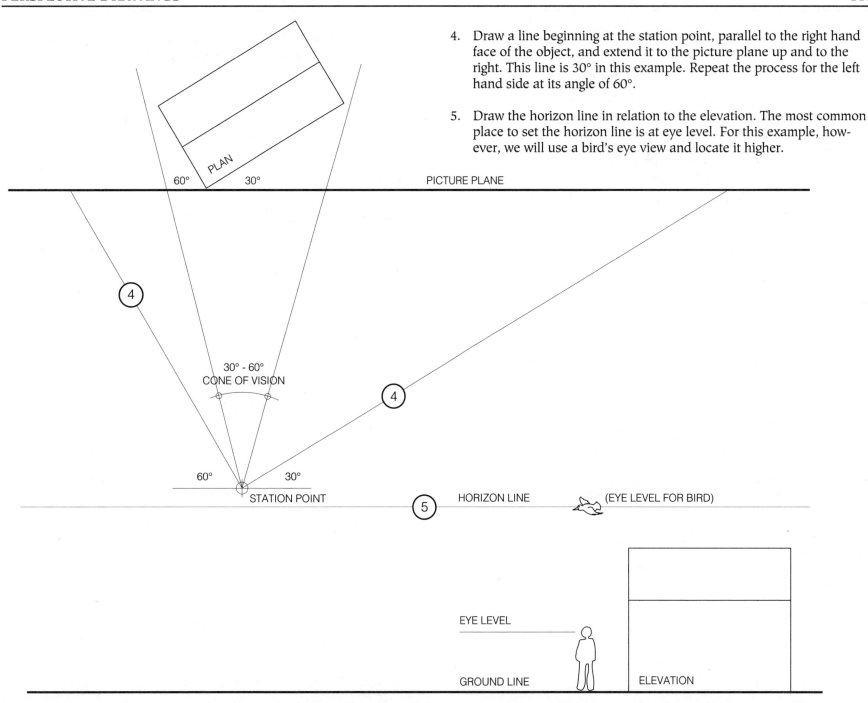

PLAN

60° 30°

PICTURE PLANE

4

30° - 60°
CONE OF VISION

4

60° 30°

STATION POINT

5 HORIZON LINE (EYE LEVEL FOR BIRD)

EYE LEVEL

GROUND LINE ELEVATION

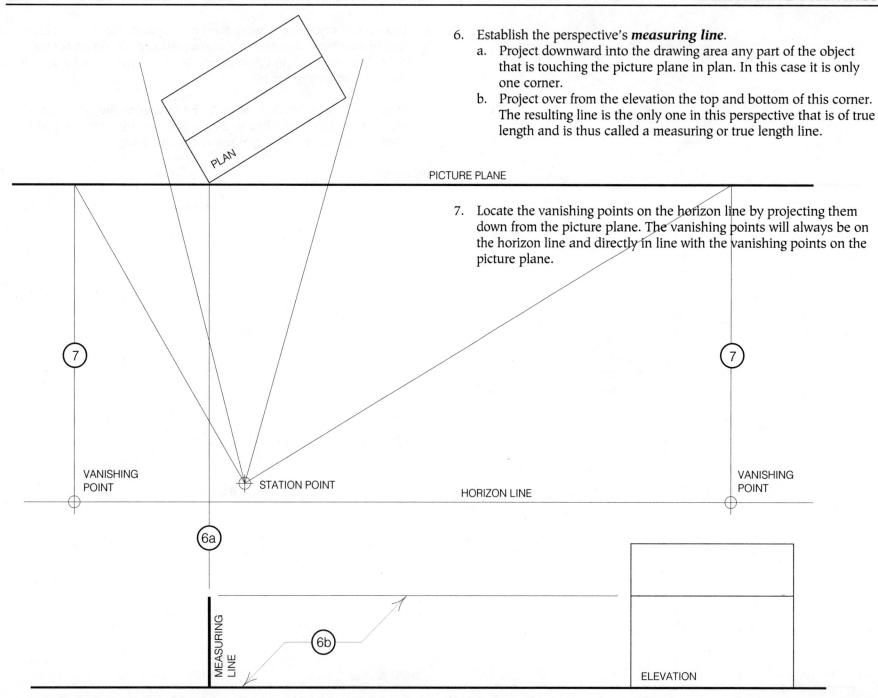

6. Establish the perspective's ***measuring line***.
   a. Project downward into the drawing area any part of the object that is touching the picture plane in plan. In this case it is only one corner.
   b. Project over from the elevation the top and bottom of this corner. The resulting line is the only one in this perspective that is of true length and is thus called a measuring or true length line.

PICTURE PLANE

7. Locate the vanishing points on the horizon line by projecting them down from the picture plane. The vanishing points will always be on the horizon line and directly in line with the vanishing points on the picture plane.

PLAN

⑦        ⑦

VANISHING
POINT                    STATION POINT                          VANISHING
                                                                POINT

HORIZON LINE

6a

MEASURING
LINE                         6b

                                                ELEVATION

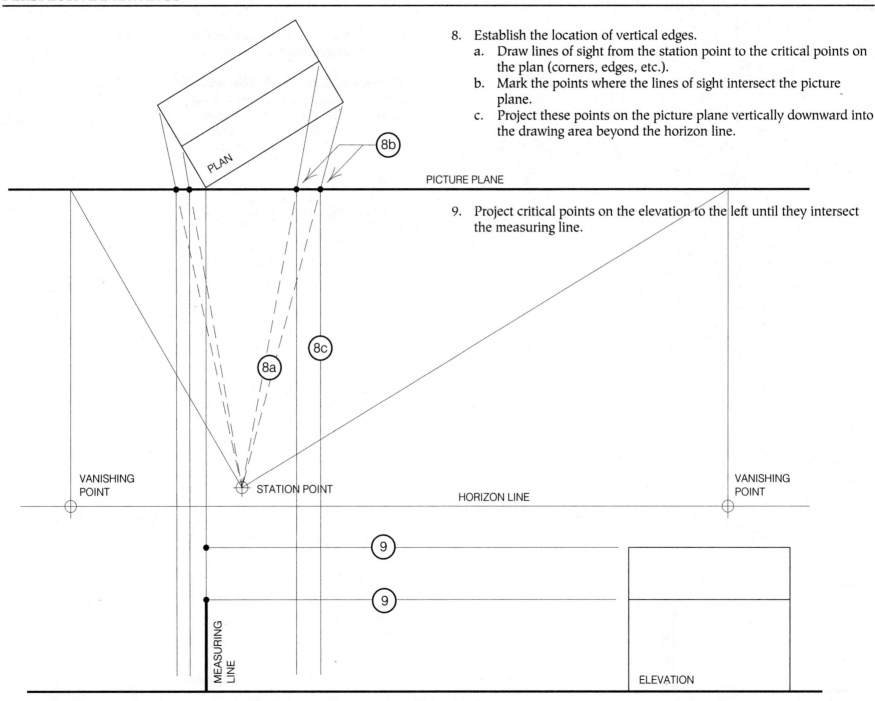

8. Establish the location of vertical edges.
   a. Draw lines of sight from the station point to the critical points on the plan (corners, edges, etc.).
   b. Mark the points where the lines of sight intersect the picture plane.
   c. Project these points on the picture plane vertically downward into the drawing area beyond the horizon line.

9. Project critical points on the elevation to the left until they intersect the measuring line.

PICTURE PLANE

PLAN

VANISHING POINT

STATION POINT

HORIZON LINE

VANISHING POINT

MEASURING LINE

ELEVATION

10. From the critical points on the measuring line, project lines back to both vanishing points.

11. Mark points of intersection between the lines drawn in steps 8c and 10. These points establish the outline of the perspective drawing.

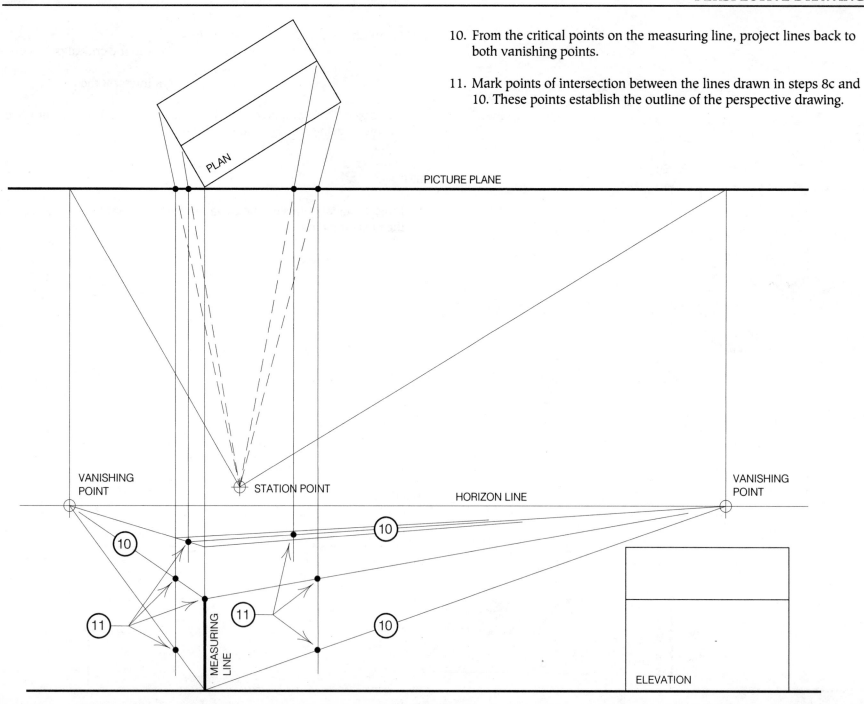

12. Complete the perspective by connecting the points identified in step 11.

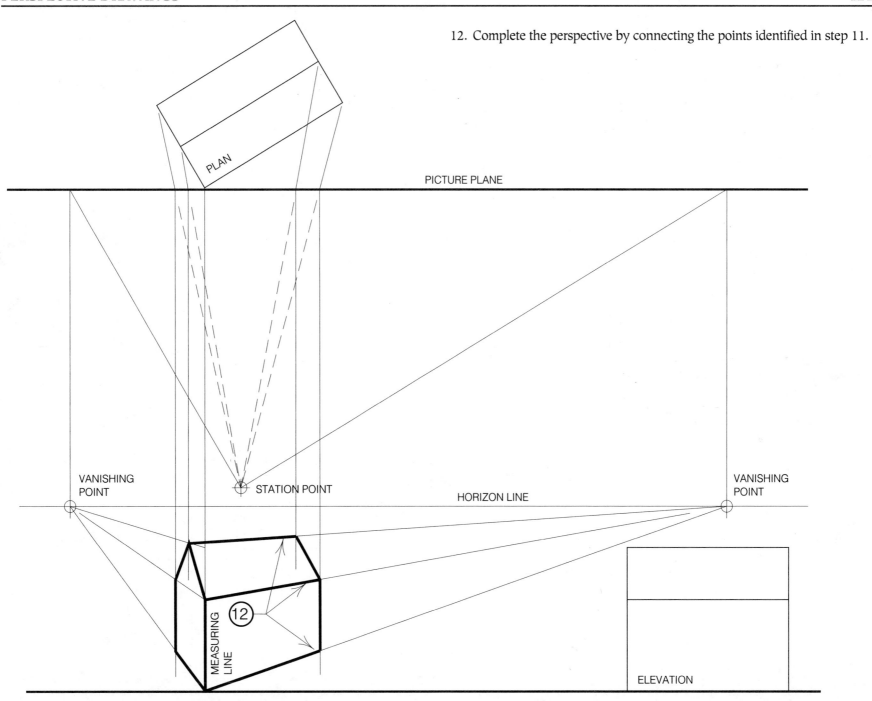

PLAN

PICTURE PLANE

VANISHING POINT

STATION POINT

HORIZON LINE

VANISHING POINT

MEASURING LINE

⑫

ELEVATION

# Key Terms

paraline drawing
cavalier oblique
general oblique
cabinet oblique
isometric
dimetric
trimetric
vanishing point
station point
cone of vision
one-point perspective
two-point perspective
three-point perspective
measuring line

# Practice Exercise

Create oblique, axonometric, one-point, and two-point perspective drawings for the following objects. The numbers indicate proportions. Select a scale to fit on the size drafting media you normally use.

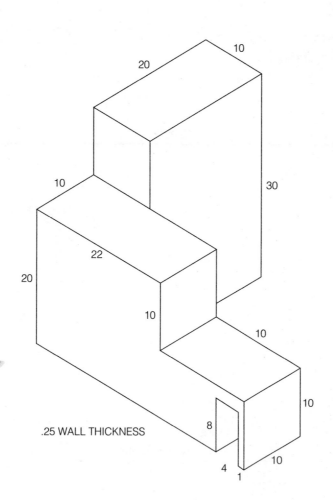

.25 WALL THICKNESS

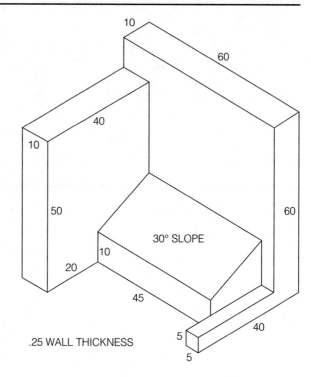

.25 WALL THICKNESS

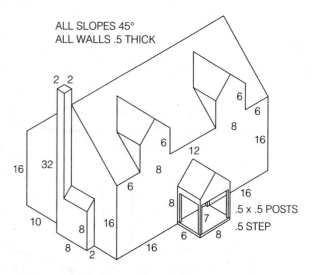

ALL SLOPES 45°
ALL WALLS .5 THICK

.5 x .5 POSTS

.5 STEP

# 10

# SHADES AND SHADOWS

## OBJECTIVES

*By the end of this chapter students should be able to:*
Construct and draw shades and shadows for
- plans and elevations
- paraline drawings
- perspective drawings

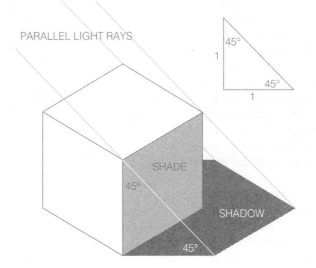

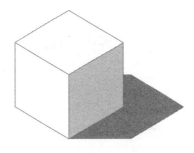

RENDERED WITH FILLS

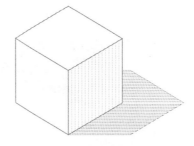

RENDERED WITH LINES

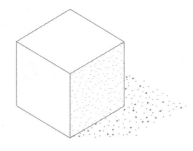

RENDERED WITH PATTERNS

# LIGHT, SHADES, AND SHADOWS

Thus far in this text we have been looking at objects without concern for light. They have been drawn as flat objects with the only illusion to depth being provided by variations in line weight.

In preliminary drawings where the purpose is to describe the character of a design rather than the technology, the use of shades and shadows is a great help in adding more realism. In this way the drawings look better, are easier to understand, and can show the effect of the sun at any given time.

Shades and shadows can be applied to two-dimensional orthographic drawings as well as three-dimensional paraline and perspective drawings. For 2-D drawings the third dimension can all but become part of the drawing where it previously did not exist at all.

Most drafters define **shade** as a surface that is receiving no direct light. The shape cast on the ground or other surface from an object is the definition of a **shadow**. Thus if you were standing with the sun at your back, the front of your body would be in shade, and the darkened outline of your body on the ground would be your shadow.

There are several guiding principles to shades and shadows that apply to all applications. They tend to make construction quite a bit easier. They are illustrated at right.

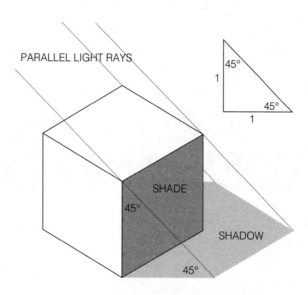

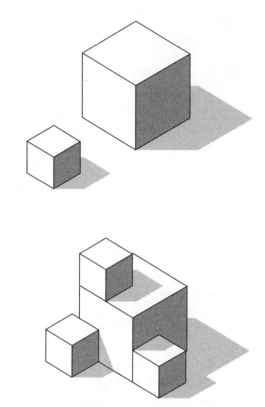

*The sun's rays are assumed to be parallel for the sake of drawing. This is not much of a compromise given the distance between the sun and earth.*

*The intensity of the sun is left up to the drafter. That is, the darkness of the shades and shadows can range from black to very light.*

*For most situations the sun is assumed to be at a 45° angle to the ground.*

*Shadows of complex objects are the sum of the shadows for the basic shapes from which the object is made.*

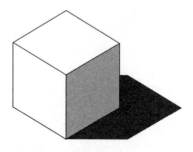

RENDERED WITH FILLS

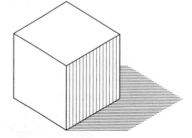

RENDERED WITH LINES

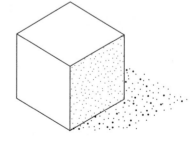

RENDERED WITH PATTERNS

*Above is an illustration of variations in the way shades and shadows can be rendered.*

## Orthographic Drawings

For orthographic drawings it is a help to construct shades and shadows with plans and elevations or sections. The sun can be assumed to come from any angle, but usually from behind the viewer. The geometry involved is more descriptive than words; hence the illustrations that follow explain the basics of shades and shadows.

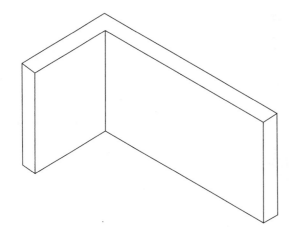

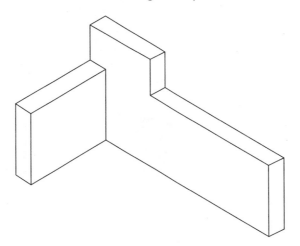

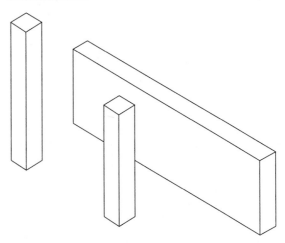

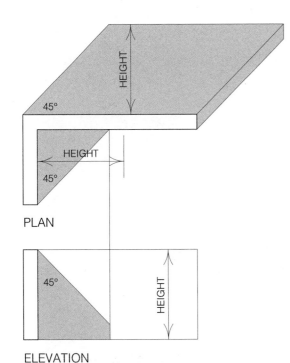

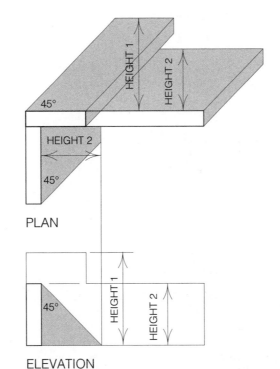

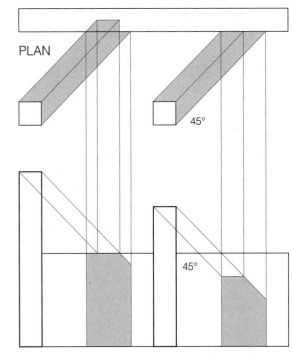

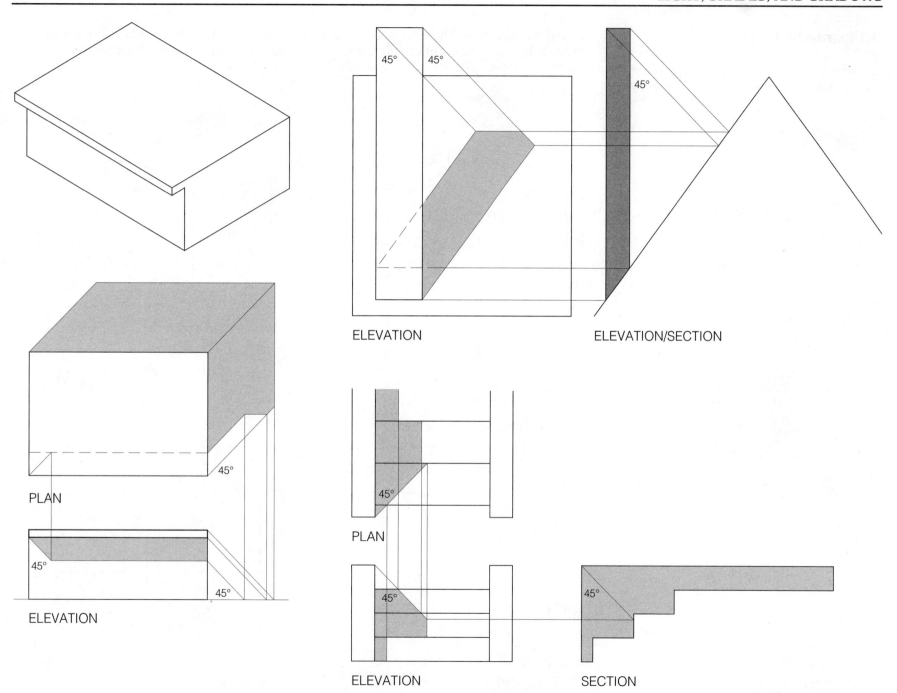

ELEVATION

ELEVATION/SECTION

PLAN

ELEVATION

PLAN

ELEVATION

SECTION

## Oblique Drawings

Oblique drawings already show three dimensions, so shades and shadows make them look all the more realistic. They are no more difficult to construct than on orthographic drawings, but once again they are more easily explained by geometry than by words.

Because oblique drawings have one face of the object square with the picture plane, the sun is usually assumed to come from behind the viewer at a 15° angle. This avoids having the shadow align with the base edge of the object or one of its sides, which is usually set at 30°, 45°, or 60°.

You will also notice that horizontal edges of the object that are parallel to the picture plane cast horizontal shadow lines. Horizontal edges oblique to the picture plane cast their shadows parallel to their respective edges.

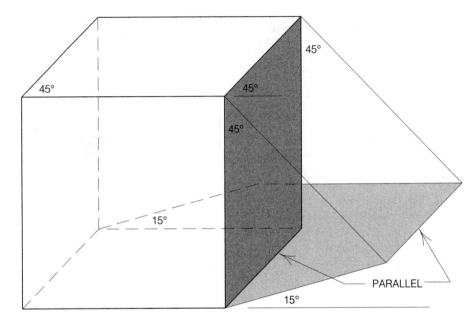

OBLIQUE VIEW AT 45° ANGLE

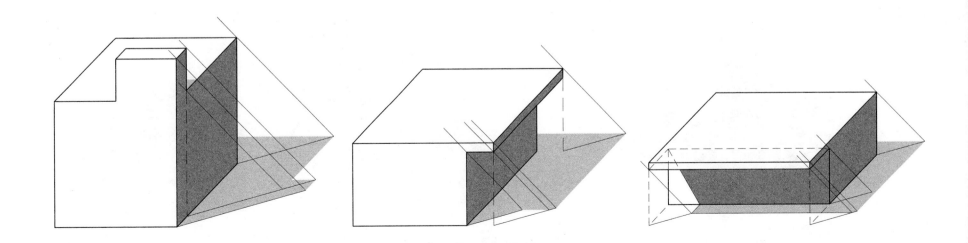

## Axonometric Drawings

Shades and shadows for axonometric drawings are very similar to those for oblique drawings. One might even conclude that they are easier.

With axonometric drawings, since no face of the object is parallel to the picture plane, the most common convention is to assume that the sun is coming from the immediate left at a 45° angle. This makes construction of the shades and shadows very simple.

Vertical edges of the object cast shadows that are horizontal lines. The other edges of the shadow are parallel to the edges of the object.

An alternate method is to cast the angle of the shadow at 15° behind the viewer, the same as oblique drawings. It is a bit more time consuming, but this method avoids the ambiguity of shadows aligning with objects, as seen on the dimetric example below.

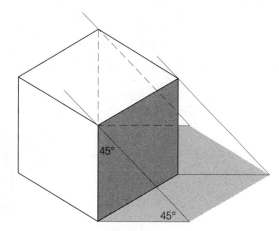

SUN DIRECTLY FROM LEFT

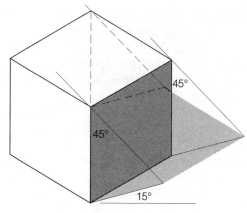

SUN FROM 15° BEHIND VIEWER

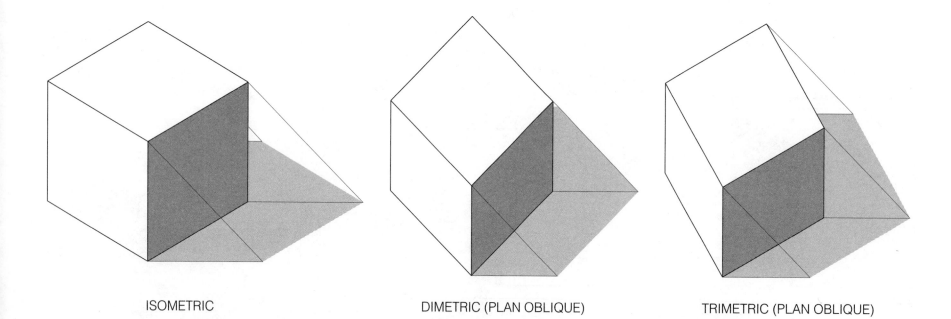

ISOMETRIC       DIMETRIC (PLAN OBLIQUE)       TRIMETRIC (PLAN OBLIQUE)

## Perspective Drawings

Shades and shadows for perspective drawings are constructed similarly to axonometric drawings. The sun is assumed to come from the immediate left or, if desired, at 15° behind and to the left. Lengths of shadows from vertical edges are determined by 45° angles as in axonometric drawings, but horizontal edges are drawn differently.

In perspective drawings the horizontal edges converge to one set of vanishing points while the shadows they create go to their own set of vanishing points. Although accurate, this drawing is more complicated to construct. Many drafters prefer an alternate method, which is illustrated here. In this example the shadows are drawn to converge to the same set of vanishing points as the edges of the object.

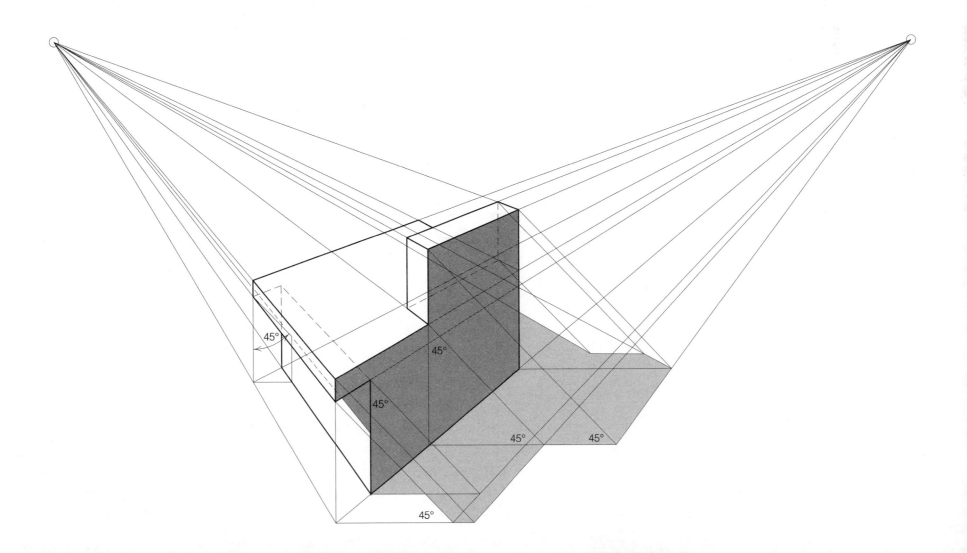

# KEY TERMS

shade
shadow

# PRACTICE EXERCISE

Redraw the objects on this page in the following views:
- plan and elevation
- oblique
- axonometric
- perspective

Then construct and draw shades and shadows for each. Numbers represent dimensions in feet. Drawings may be done at any scale.

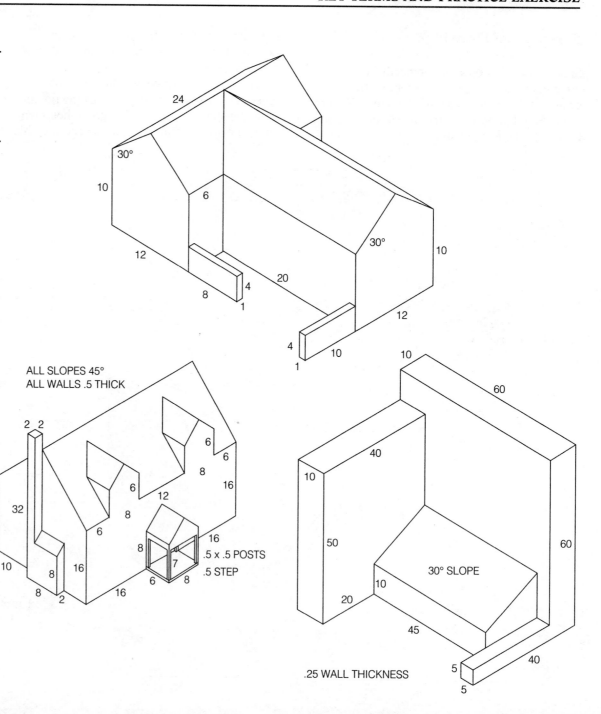

# 11

# PRE-DESIGN DRAWINGS

- - - 87 - - - CONTOUR LINE

+ 67.78    67.78    67.78    SPOT ELEVATION

——————— PROPERTY LINE    67.78

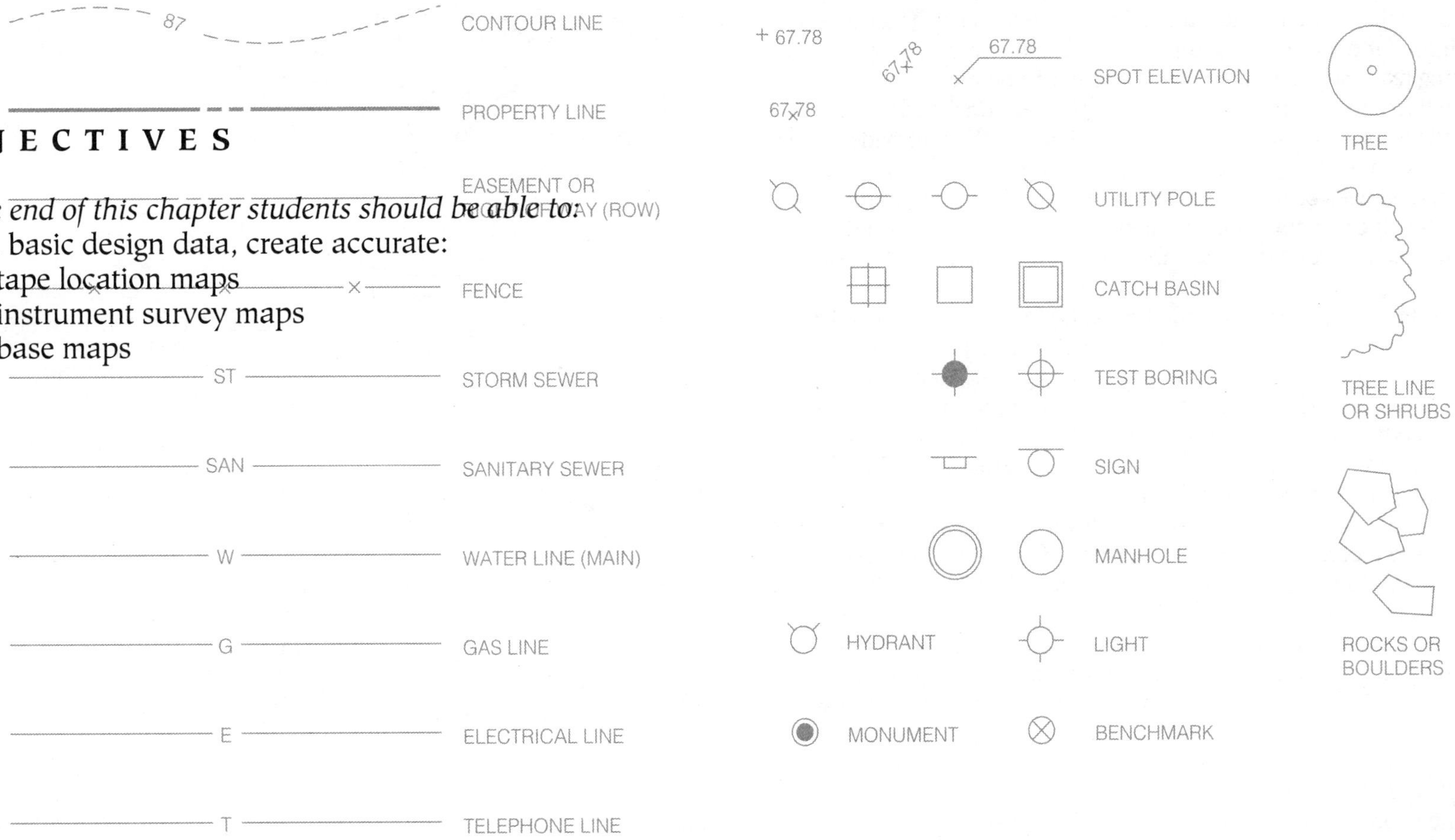

## OBJECTIVES

*By the end of this chapter students should be able to:*
Given basic design data, create accurate:

EASEMENT OR RIGHT-OF-WAY (ROW)

- tape location maps    ×    FENCE
- instrument survey maps
- base maps

TREE

—— ST —— STORM SEWER    TEST BORING

—— SAN —— SANITARY SEWER    SIGN    TREE LINE OR SHRUBS

—— W —— WATER LINE (MAIN)    MANHOLE

—— G —— GAS LINE    HYDRANT    LIGHT    ROCKS OR BOULDERS

—— E —— ELECTRICAL LINE    MONUMENT    BENCHMARK

—— T —— TELEPHONE LINE

UTILITY POLE

CATCH BASIN

Before a project can be designed, let alone built, all parties involved need to know exactly where the property lines are and what features are found within them. A site survey is conducted to investigate what is above and below the surface. Generally a civil engineer can do this survey. A surveyor specializes in the surface measurements, and a soil testing firm specializes in subsurface surveys. In many instances, a large engineering firm can provide this expertise as well.

## PLAT MAPS

In the United States and most other countries there is a system of surveying and record keeping for all of the land within a country's boundaries. In the case of the United States, the federal Government has assumed responsibility for dividing the country into measurable parcels of land and establishing fixed points from which property can be measured.

These fixed points are called *monuments* and consist of brass headed spikes usually encased in concrete. The idea is to provide permanent markers from which surveyors may begin their surveys. This system avoids discrepancies or property disputes when adjacent properties are measured by different surveyors at different times. If property measurements originate from one of these monuments, then there is little room for error.

After individual properties (called *parcels*) have been surveyed and before any structures are placed on them, their records are filed at the local courthouse or other government office designated to maintain records. The survey, which consists of a legal description (words) and a drawing, called a *plat*, form a record of the property. Much of this work was originally done as the country grew and was settled. The illustration on the facing page shows a typical plat map.

Today, every time property is physically altered or is sold, the change is added to the record. Thus for a given piece of property there exists a physical description and a historical record that dates back to the first time the property was measured.

As time goes on, large parcels are subdivided into smaller ones. The typical situation is a farm that is sold to a developer for construction of a housing project. The large parcel is divided into many individual parcels, or *lots*, where single dwelling units are built. The result is that new plats are established and filed. Thus the process continues, and the records in county office buildings grow larger.

For the situation where property is bought and sold, the lending institution that finances the transaction wants to protect its interest in the property. They want to be sure that everything is as the borrower says it is and as it is referred to in the purchase offer for the land. To do this, they want to know where the property is, what size it is, what is on it, who owns it, etc. To get this information they require the purchaser to obtain a map from a surveyor or title company that includes this type of information. Basically there are two types: tape location maps and instrument survey maps.

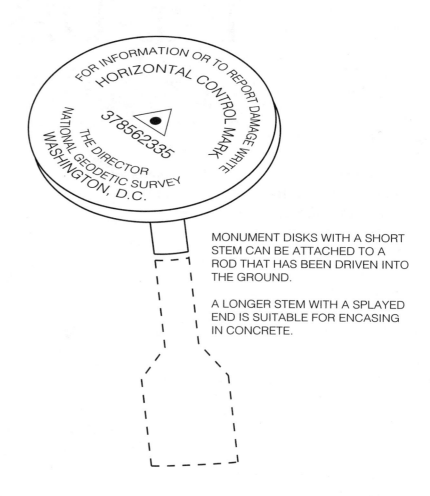

MONUMENT DISKS WITH A SHORT STEM CAN BE ATTACHED TO A ROD THAT HAS BEEN DRIVEN INTO THE GROUND.

A LONGER STEM WITH A SPLAYED END IS SUITABLE FOR ENCASING IN CONCRETE.

*National Geodetic Survey markers, known as monuments, are found throughout the United States. Surveyors use them as starting points for their work.*

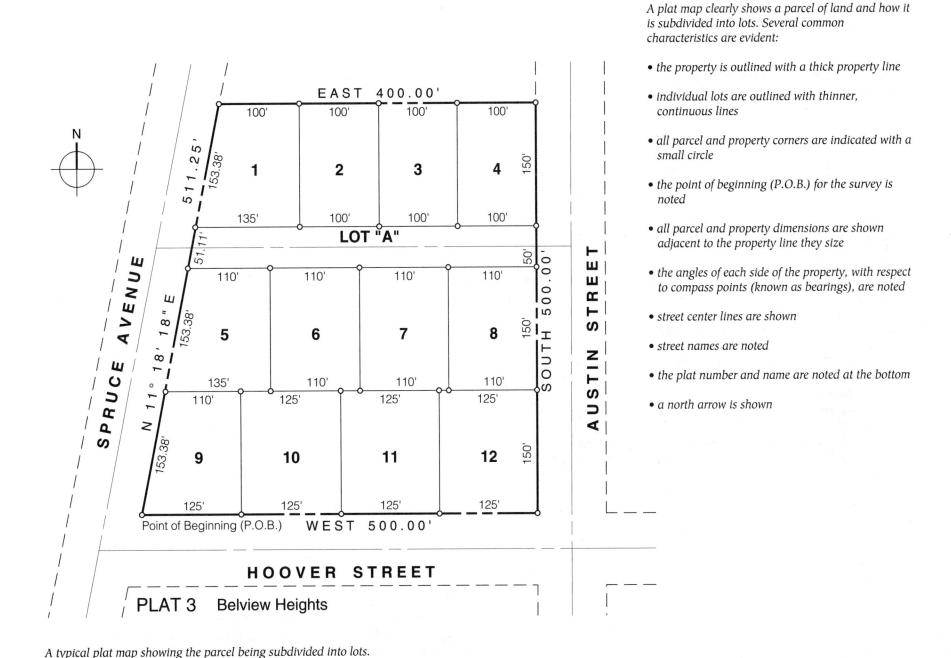

PLAT 3   Belview Heights

*A plat map clearly shows a parcel of land and how it is subdivided into lots. Several common characteristics are evident:*

- *the property is outlined with a thick property line*

- *individual lots are outlined with thinner, continuous lines*

- *all parcel and property corners are indicated with a small circle*

- *the point of beginning (P.O.B.) for the survey is noted*

- *all parcel and property dimensions are shown adjacent to the property line they size*

- *the angles of each side of the property, with respect to compass points (known as bearings), are noted*

- *street center lines are shown*

- *street names are noted*

- *the plat number and name are noted at the bottom*

- *a north arrow is shown*

*A typical plat map showing the parcel being subdivided into lots.*

# TAPE LOCATION MAPS

The easiest form of site drawing is called a *tape location map* where, as the name implies, the surveyor uses a steel tape measure to locate site features such as buildings. Shown at right is a typical tape location map. There is some important verbal description included on the map such as who it was done for, the exact location of the property (with reference to maps already on file), and how the survey was done.

The main purpose of the drawing on this document is to show the property with the building on it. It is important to realize that the boundary lines have not been measured but have been taken from the county records. As a matter of fact, at the top of the sheet a box has been checked stating that monuments were not used in preparing the map. Again at the bottom is the statement, "This map is not intended or represented to be a land or property line survey. No corners were set. Do not use for establishing fence or building lines."

Thus a tape location map is considered not accurate enough to base construction work on but adequate for its intended purpose. What it offers is a quick and easy way to satisfy the needs of a lending institution when it provides money for the sale of a property. When drawing this map the drafter should incorporate the following charac-teristics:

- the property lines and street line should be drawn
- the length of the property lines should be noted adjacent to the property lines using large lettering
- the street name should be noted in large lettering with its width noted in smaller lettering
- the building should be shown in its proper location
- the building should be dimensioned back from the street line and from at least one side lot line
- each face of the building should be dimensioned
- the construction type of the building and the height should be given in stories or feet
- roof lines at overhangs should be drawn as hidden lines and noted
- a north arrow and scale should be included
- all dimensions should be measured and given using an engineer's tape and scale

STANDARD TAPE LOCATION MAP

**ROBERT M. QUINN**
Professional Land Surveyor

227 Culver Road    Rochester, NY 14607    (716) 555-1276

City / Village
PITTSFORD Town
MONROE County

For RUSTON, LEONARD & ELIZABETH ___ Parcel at ___
Street 22 CANDLEWOOD DRIVE  Lot No. 11  Subdivision BEL VIEW HEIGHTS SECTION 1
Reference Data: Liber 151 of Maps, Page 63; Liber 3506 of Deeds, Page 501
Showing FRAME 2 story dwelling; garage attached. Monuments used ☐ Yes ☑ No
Distance as shown from W & N property line actually measured. Main front wall is an apparent uniform set-back line.

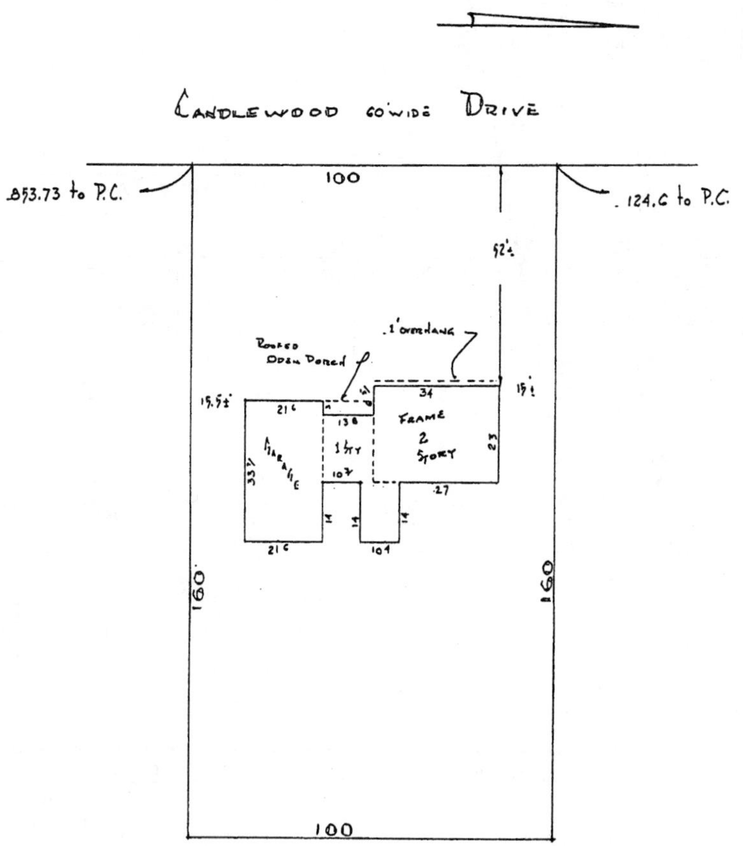

SCALE 1" = 30'
Remarks: This information is for JOHN A. BUYOR

This map is not intended or represented to be a land or property line survey. No corners were set. Do not use for establishing fence or building lines.

Dated APRIL 12, 1993          Signed Robert M. Quinn

MEMBER GENESEE VALLEY LAND SURVEYORS ASSN.          PROFESSIONAL LAND SURVEYOR NO. 15877

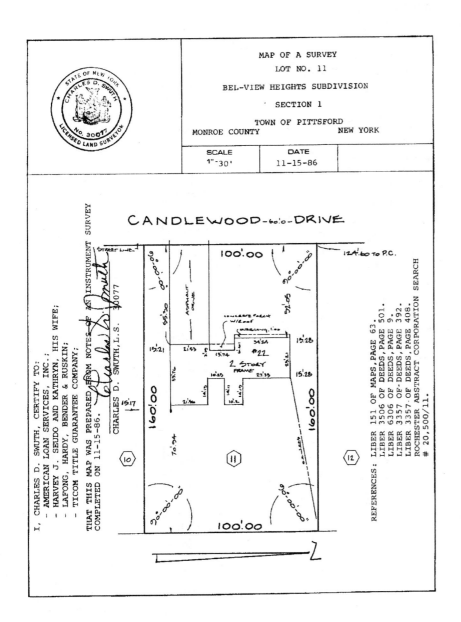

# INSTRUMENT SURVEY MAPS

Surveyors can produce maps that are more accurate than tape location maps. The more sophisticated method is the ***instrument survey map***, which uses precision equipment to define the physical properties of the site. In many cases today, instrument survey maps are preferred over tape location maps by lending institutions when a property is transferred.

An instrument survey is in essence a re-survey of the property, which allows the surveyor to verify what is already on record and note any modifications. In turn, it becomes the latest addition to the record of that piece of property.

The surveyor actually begins the survey at a monument or benchmark and re-establishes the property boundary. A ***benchmark*** is merely any fixed point of known height from which other surveys may commence. The top of a manhole cover or a point on a foundation wall could serve as a benchmark if it has been previously set up as one. The importance of an instrument survey is that the equipment used is capable of measuring much more accurately than a person can by hand.

A look at the map on this page allows one to see the additional and much more accurate information provided on an instrument survey map. This is the same property shown on the facing page, but notice that the dimensions are given to the 1/100th of a foot, the angles of the corners are given, the driveway is shown, the house is dimensioned from all four property lines, and even the above ground utility wires are shown.

The accuracy of this document is also clearly stated . . . "this map was prepared from notes of an instrument survey completed on 11-15-86." It has been certified by the licensed land surveyor who prepared the document and also includes a print of his licensing stamp.

The key characteristics of this type of drawing, which differ from those of the tape location map, are as follows:
- dimensions are given to the hundredths decimal place
- the building is drawn and dimensioned to each property line
- paved areas, utilities, and other improvements are drawn and dimensioned when appropriate
- angles of corners are given
- the drawing is sealed with a professional licensing stamp
- easements and setbacks may be noted

# BASE MAPS

When developers decide to build a project on property that they own or intend to own, they need to provide site information to the architects, engineers, and contractor that goes well beyond that found on an instrument survey map.

The information on an instrument survey map is certainly a good starting point, but the designers and builder need to know other things. They need to know the slope of the land, or **topography**, where underground utilities are, where vegetation is, and if any water or rock formations are present.

To obtain this information, a developer, or an architect or engineer acting on the developer's behalf, hires a surveyor, civil engineer, or soils engineer to do a survey. This survey is usually comprised of three parts:

- soils survey
- land survey
- topographic survey

A base map is essentially a combination of the maps produced from these surveys. A soils survey shows the location of the soil samples taken; a land survey produces the plat or instrument survey map; and the topographic survey produces a topographic or "topo" map.

## Soils Maps

To get an understanding of the subsurface conditions of a site, samples are taken of the soil and tests performed to determine the properties of the soil. Subsurface information may be obtained by any one of several methods, including soil borings (drilling holes), test pits (digging holes), and electronic or sonic measuring (like sonar that bounces waves off of the layers of soil).

A report is generated showing the results of these tests, part of which is a map showing the exact location of each test. This is a very simple map, but it is critical in matching where each type of soil may be found on the site.

TEST BORING #3

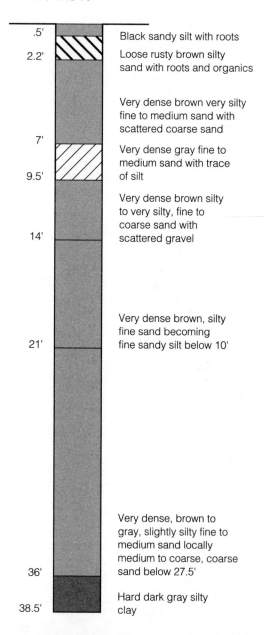

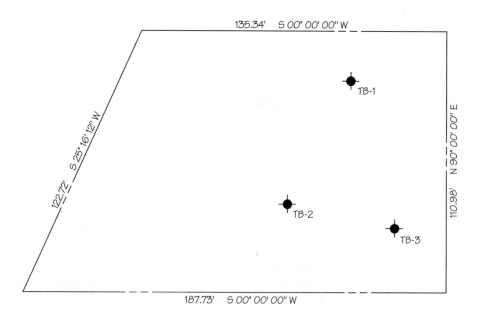

## Survey Maps

A land survey is in essence an instrument survey of the property that establishes the property lines very accurately as well as existing improvements.

One difference between an instrument survey done for a lending institution and the one done for a developer and consultants is the way in which the relationship between property lines is defined. The instrument survey map shown previously gives the angle between property lines, while a base map almost always gives each line a *bearing*. A bearing is the angle a line makes with one of the *cardinal points* (exactly north, south, east, or west) of the compass.

Suppose a property line runs basically north and south but tilts towards the west at 30°. Its bearing would be N 30° W, meaning the angle between this line and true north is 30°, starting from the north and moving towards the west. Another way of expressing the same line is S 30° E. In general we always define the angle in terms of beginning at north or south. Thus W 60° N and E 60° S would be incorrect definitions of this line.

Angles can be measured more accurately than just degrees however. One degree is divided into 60 equal parts called *minutes*, and each minute is divided into 60 equal parts called *seconds*. It is the same terminology and concept we use for measuring time, but the units of measure are quite different. Thus the bearing of a property line measured to the utmost accuracy might read "north 15 degrees, 27 minutes, 45 seconds west" or abbreviated N 15° 27' 45" W. The symbols for minutes and seconds are none other than the familiar feet and inches symbols we are all familiar with.

Property measurements using the bearing method need to start at one specific corner of the property called the *point of beginning* or P.O.B. for short. This point is referenced back to a benchmark or monument. From it one side is established, the next side continues from that, and so on until the last side returns to the P.O.B.

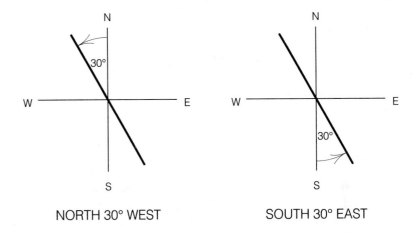

NORTH 30° WEST          SOUTH 30° EAST

*Bearings are easily established by starting at a cardinal point and then moving a specified number of degrees toward another cardinal point.*

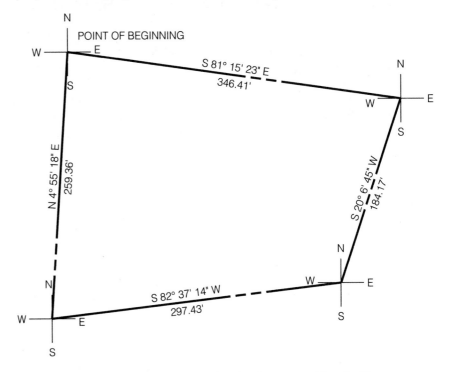

*A boundary survey is done by giving each side a bearing and length. The surveyor starts at the point of beginning and continues around the property until arriving back there, in this case, moving around the property clockwise.*

## Topographic Maps

We can now turn our attention to the information provided by the topographic survey and how it is added to the base map. Property lines are fixed and define a parcel in the X and Y directions. In the Z direction, the height of the land varies. Some sites are quite flat; others are extremely hilly. In either case it is important for the designers to know what they are dealing with before they can put a building or parking lot on the site.

A map from a topographic survey utilizes **contour lines** to indicate slope on the land. A contour line is the connection of all points on a site that are the same height. A contour line is the line created when a horizontal section is taken through the site at a specific height.

On a contour map, there are a series of lines with numbers. Each line represents a "slab" of earth at the height (**elevation** in engineering terms) indicated. Usually the number represents the height in feet above sea level; in many cases, however, it may simply be the number as related to a benchmark.

For most topo maps, contour lines are drawn for each foot of elevation. For large parcels or steeply sloped sites it would be impossible to show every foot of elevation, so the drafter may show every 2, 5, or even 10 feet of elevation.

Contour lines cannot cross since such an intersection would mean a spot of land exists at two different elevations. They can coincide, however, indicating a shear, vertical surface such as a cliff or a retaining wall. Common configurations are shown below.

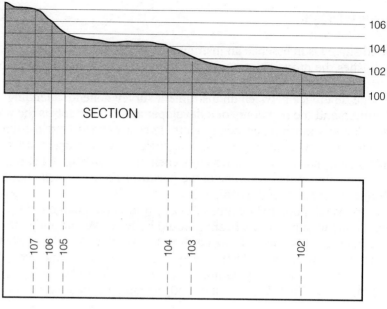

SECTION

PLAN

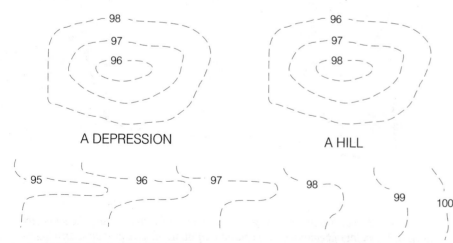

A DEPRESSION

A HILL

A CREEK OR DITCH FORMING, "V" SHAPES ALWAYS POINT UP HILL

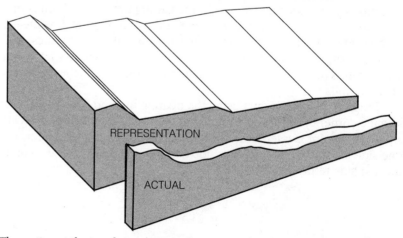

REPRESENTATION

ACTUAL

*The section at the top shows various elevations or heights indicated by the numbers 106, 104, etc. The points of intersection between these heights and the land can be transferred to the plan very easily and noted there. A trained eye can read the numbers and realize that they represent land sloping from left to right, steeply at first and then more gradually.*

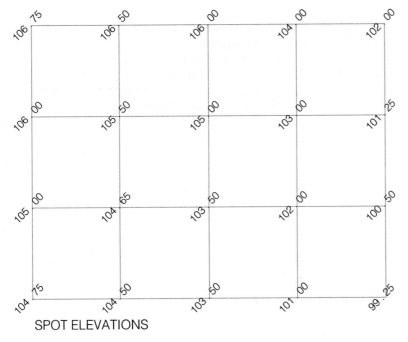

SPOT ELEVATIONS

The drawings on this page show how contour maps are generated. As shown in the top drawing, heights that the surveyor measured are entered. Notice how this is done on a grid pattern. It is rare that the point of measurement (known as a *spot elevation*) is a whole number. To ascertain where the whole number falls between spot elevations, the drafter must *interpolate* or estimate where the whole number elevation falls.

The two spot elevations at the upper right are 104 and 102 respectively. The drafter can interpolate that elevation line 103 passes equidistant between them. The drafter needs to find the whole number elevation points between each pair of spot elevations. This is shown in the lower left hand drawing. The contour lines are then drawn by connecting the whole number spot elevations, shown in the final drawing below.

Existing contour lines are shown as long broken lines of thin line weight. It is common to emphasize each fifth foot as a medium line for easier reading. The lines are drawn completely across the property, and their elevations are entered at each end. Long lines may be broken as many times as necessary, and the elevations placed within them.

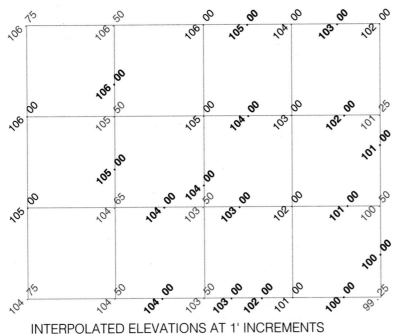

INTERPOLATED ELEVATIONS AT 1' INCREMENTS

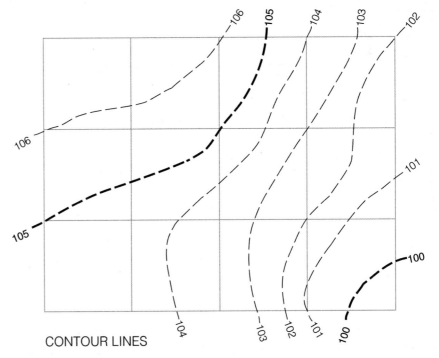

CONTOUR LINES

Aside from contour lines, the topo map also shows other natural features such as individual trees, heavily treed areas, creeks, ponds, and rock outcroppings. For individual trees, size and variety need to be noted. The size is given in *caliper*, the diameter of the trunk 1 foot above the ground. All the other natural features are shown and noted in a less specific way. Typical symbols and commonly used techniques are shown below.

The so called "base map" explained in this section is often a composite drawing, but it does not always occur as a single drawing. Often, all of the information does not fit onto one drawing, or perhaps one firm performs the topographic survey and another firm the land survey. Whatever the case, all the information needs to be gathered, whether it be on one map or several. The drawing on the facing page is a composite base map.

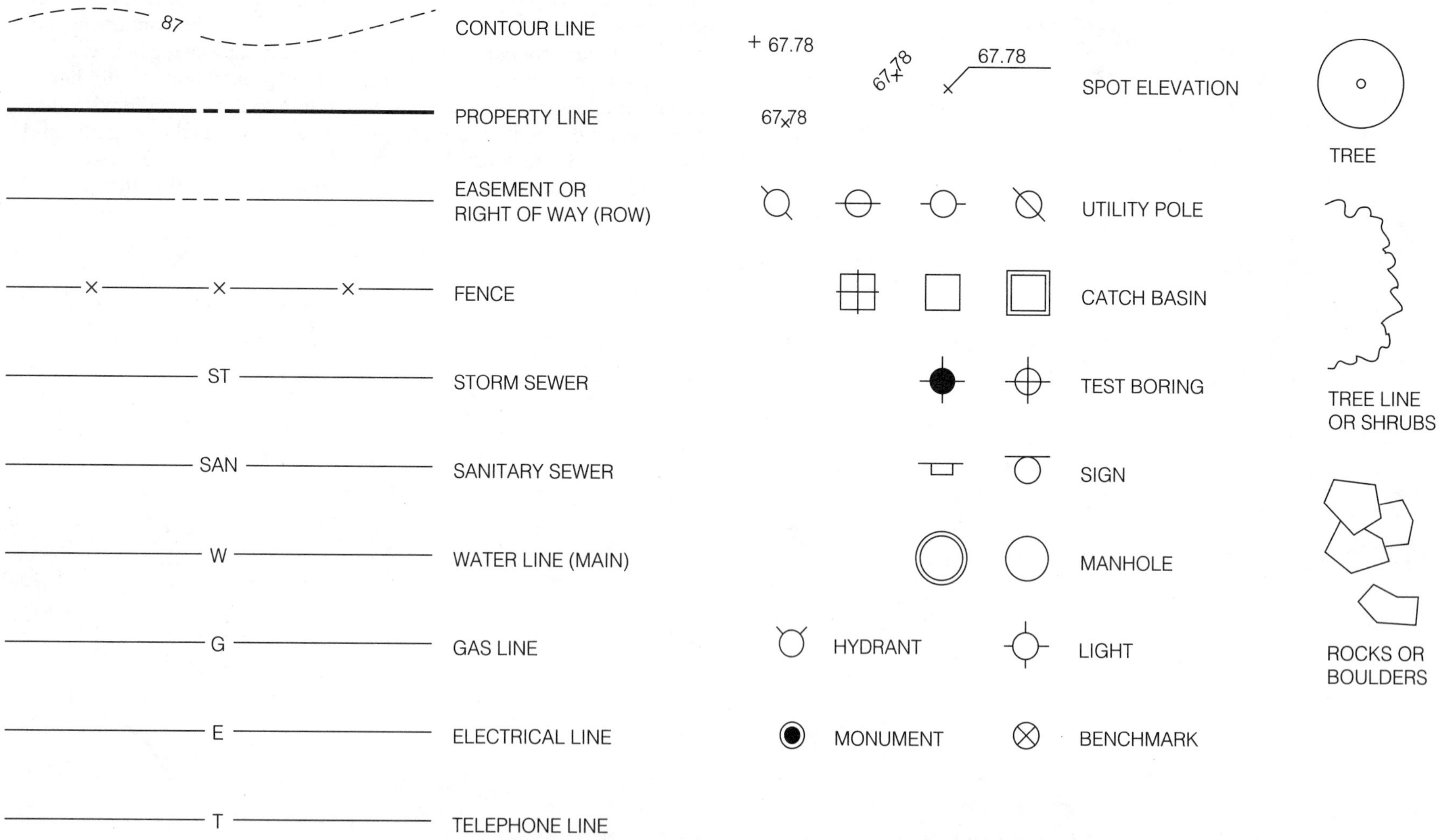

87 CONTOUR LINE

PROPERTY LINE

EASEMENT OR RIGHT OF WAY (ROW)

—x—x—x— FENCE

— ST — STORM SEWER

— SAN — SANITARY SEWER

— W — WATER LINE (MAIN)

— G — GAS LINE

— E — ELECTRICAL LINE

— T — TELEPHONE LINE

+ 67.78

67.78

67.78

67.78    SPOT ELEVATION

UTILITY POLE

CATCH BASIN

TEST BORING

SIGN

MANHOLE

HYDRANT    LIGHT

MONUMENT    BENCHMARK

TREE

TREE LINE OR SHRUBS

ROCKS OR BOULDERS

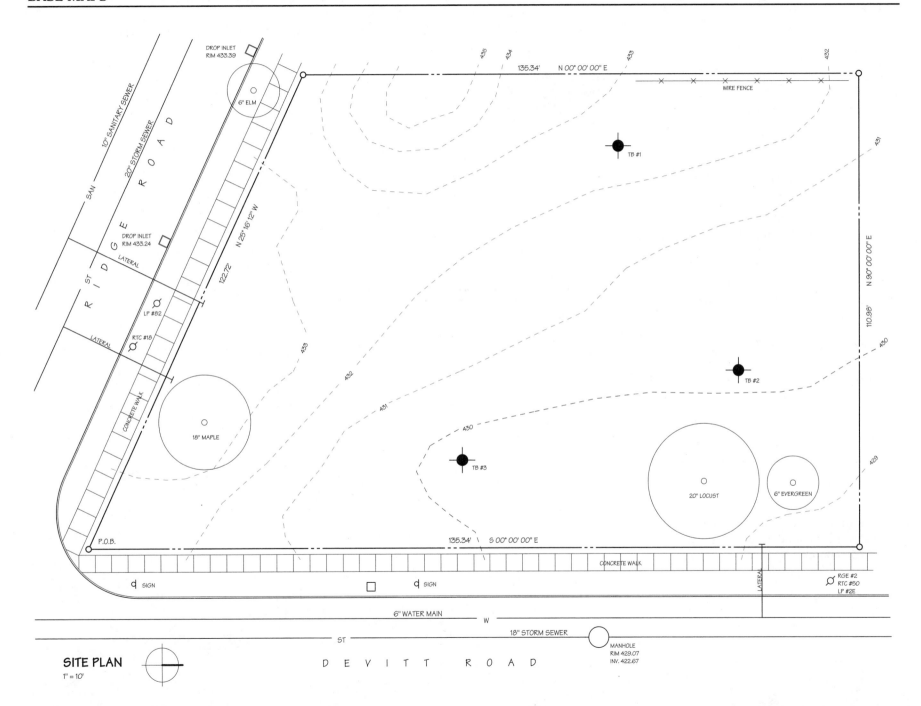

DROP INLET
RIM 433.39

6" ELM

DROP INLET
RIM 433.24

LP #82

RTC #18

18" MAPLE

P.O.B.

CONCRETE WALK

10" SANITARY SEWER

20" STORM SEWER

SAN

R    I    D    G    E        R    O    A    D

LATERAL

LATERAL

N 25° 16' 12" W

122.72'

135.34'    N 00° 00' 00" E

WIRE FENCE

435   434   433   432

431

430

429

TB #1

TB #2

TB #3

20" LOCUST

6" EVERGREEN

N 90° 00' 00" E

110.98'

433

432

431

430

135.34'    S 00° 00' 00" E

CONCRETE WALK

SIGN                    SIGN

LATERAL

RGE #2
RTC #50
LP #2E

6" WATER MAIN

W

ST

18" STORM SEWER

MANHOLE
RIM 429.07
INV. 422.67

**SITE PLAN**
1" = 10'

D    E    V    I    T    T        R    O    A    D

# KEY TERMS

monument
parcel
plat
lot
tape location map

instrument survey map
benchmark
topography
bearing
cardinal points
minutes
seconds

point of beginning
contour line
elevation
spot elevation
interpolate
caliper

# PRACTICE EXERCISES

1. Draw a tape location or instrument survey map for the following piece of property. Part of the SW 1/4 of Section 18, Township 45 North, Range 15 West, of the 6th Prime Meridian of the Official Plat of Lincoln County, Utah is more particularly described as follows:

   Commencing at the SE corner of said section 18, thence 516 feet north, thence 55 feet west to the POINT OF BEGINNING, thence west a distance of 800.00 feet, thence north 13° 30' 00" east a distance of 466.48 feet, thence east a distance of 691.10 feet, thence a distance of 453.59 feet to the point of beginning, containing 7.76 acres more or less.

2. Given the survey data provided at the right, draw a base map according to standards presented in this chapter.

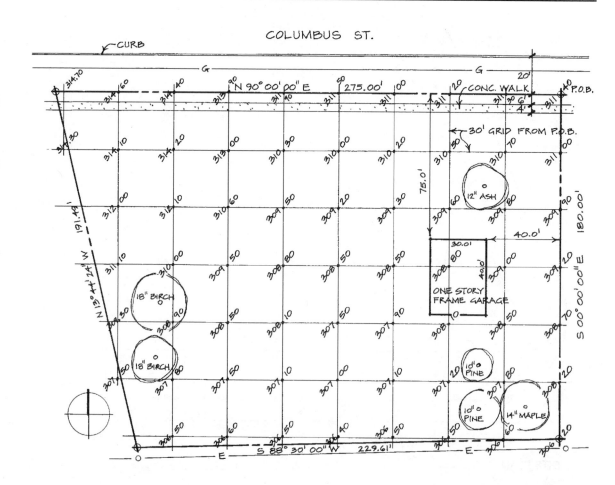

# 12

# DESIGN DRAWINGS

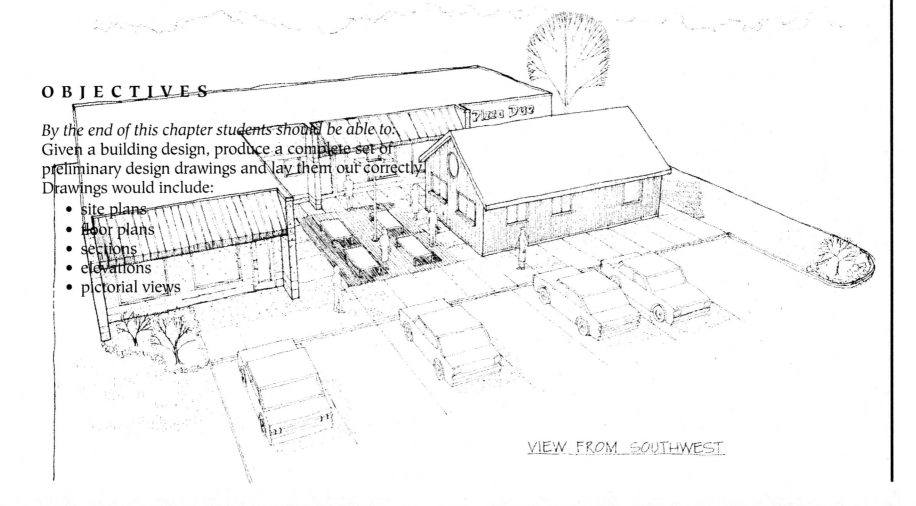

**OBJECTIVES**

*By the end of this chapter students should be able to:*
Given a building design, produce a complete set of
preliminary design drawings and lay them out correctly.
Drawings would include:

- site plans
- floor plans
- sections
- elevations
- pictorial views

VIEW FROM SOUTHWEST

# DESIGN DRAWINGS

Once existing conditions are documented on a base map, architects and, later, engineers, can begin their work. Initially the architect's job is to brainstorm and explore many possible solutions to satisfy the owner's requirements, given the realities of the site.

For very large projects, planners come up with the general layout for structures, roads, parking, and open areas on the site. For single building projects the architect is often the one who designs the site. And for projects that involve extensive site improvement (such as seating areas, walkways, planting beds, etc.), landscape architects become involved.

The early drawings produced by these individuals go by many names: **design drawings, preliminary drawings, presentation drawings, schematic design drawings, concept drawings**, etc. Because architects are responsible for the overall design process in a building construction project, we will use their terminology.

As one might recall from the first chapter, architects and engineers are most involved in the design and construction stages of a building project. Architects generally follow the guidelines of the American Institute of Architects and break down their work for these stages into five essential phases, three of which are considered part of the design process. The phases and the amount of time spent on each are as follows:

Design Stage
- schematic design               15%
- design development              20%
- contract documents             40%

Construction Stage
- bidding and/or negotiation       5%
- construction                    20%

Schematic design drawings are true design drawings because their sole purpose is to convey the design. Design development drawings are those that bridge the gap between the schematic design and the totally technical contract document drawings. Contract document drawings, also known as working drawings, depict for the builder how the design is to be constructed.

Dividing the design stage into three parts is done primarily for legal agreement purposes between the architect and owner. It helps to have discrete points of approval and payment. The design **process** is a continuous one, however, and the lines between phases are not always distinct. This is especially true of the schematic design and design development phases.

There is a great deal of variation from one office to another and from one type of project to another. Some firms have tried to combine the first two phases into one, which can be viable for certain types of projects. Other firms keep the schematic design phase much more sketchy and emphasize the design development with very formal drawings. What this says is that schematic design drawings from one office can exhibit the same characteristics as design development drawings from another office.

As an introductory text we will not attempt to definitively separate these two phases. Rather, as the title of the chapter suggests, we will treat this topic as a single one, i.e., design drawings. Only after some experience can a student begin to see more clearly the transition from pure design drawings to pure technical drawings.

Following are formal descriptions for each of the first two phases. The drawing examples that follow are again more representative of a concept than of a phase. The remaining chapters of the text will be devoted to working drawings, the drawings that constitute the largest expenditure of time for the design professions and especially the drafter.

## Schematic Design Drawings

The main objectives of the schematic design phase are for the architect to come up with a feasible solution to the owner's building needs, to present the solution to the client so that it is understandable, and to get the owner's approval of the solution.

To meet those objectives, the architect explores as many solutions as possible, making rough sketches of each. The pros and cons of each are weighed, and the strongest solution is identified.

The final scheme is refined and drawn up into a set of drawings which, although still sketchy in nature, are of presentation quality. Sometimes physical or computer models are necessary to fully explain the design.

Secondarily, the drawings must explain the design to investors when required, and to the local building officials such as the planning board, architectural review board, etc.

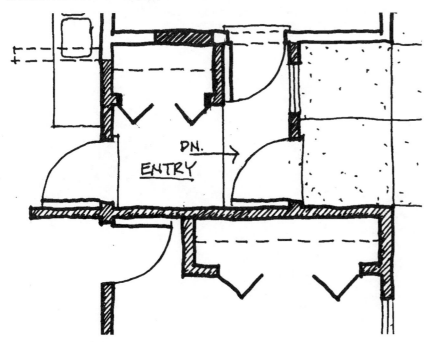

*Schematic design drawings are frequently freehand drawings that basically get the design idea across. They are not technical at all. Assumptions may be made about sizes of building components, but they are in no way final.*

## Design Development Drawings

The objectives of the design development phase are different from those of the schematic design phase. The primary purpose is to define and describe all important aspects of the project so that all that remains is the formal documentation step of construction documents. One might be able to see that these are in essence further refinements of the schematic design.

During the design development phase, the structural, mechanical, electrical, and other consultants are called in to design their portions of the project. Drawings may need to be submitted to governmental agencies for a progress check. Ultimately the client needs to approve all of the decisions made by the design team during this phase.

Experienced technicians and drafters are instrumental in the success of this phase of documentation.

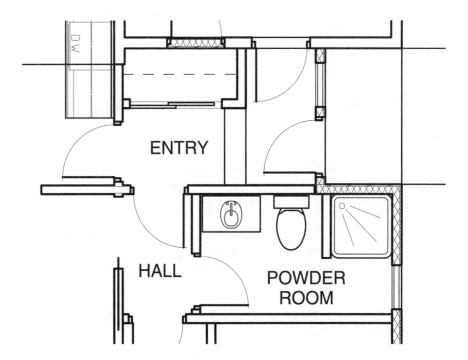

*Design development drawings may still be freehand drawings but are more commonly drawn with equipment, either manually or with CAD. They are technical in the sense that exact sizes are assigned to all components.*

# SITE PLANS

As with all design drawings, the objective of a site plan is to explain the design to the client. Realism plays an important part. The drafter needs to make the site plan look like an actual top view of a model. To do this the following techniques may be employed.

❑ The building should be emphasized by showing the roof plan and adding shadows or a heavy outline.

❑ The sun is usually assumed to come from the top left or right of the sheet rather than the south. This will make the building seem to pop up out of the page rather than fall out of it.

❑ An alternate way to make the building stand out is to shade the entire background around the building and leave the building(s) the color of the medium (usually white).

❑ The property line is the next most important feature and should be emphasized.

❑ All landscaping should be shown complete with shadows.

❑ Surface materials should be shown and noted.

❑ All site design features should be shown and noted, including driveways, parking, courtyards, sports fields, gardens, etc. The number of parking spaces should be noted.

❑ Vehicles, people, and any other objects appropriate to the project should be drawn with shadows. Keep in mind, however, that we do not want to distract from the design or hinder the client's understanding of it.

❑ Some indication of slope should be shown, reflecting both existing and proposed contours.

❑ Streets should be titled.

❑ The drawing should be titled, scaled, and should contain a north arrow.

❑ One convention for all plans is to show north to the top of the sheet. If this is not possible then it should be to the right of the sheet as shown on the facing page.

❑ Also popular in the industry is to have the direction of entry into the property and building along the bottom of the sheet. This orients the project to the viewer as if the viewer were actually approaching it.

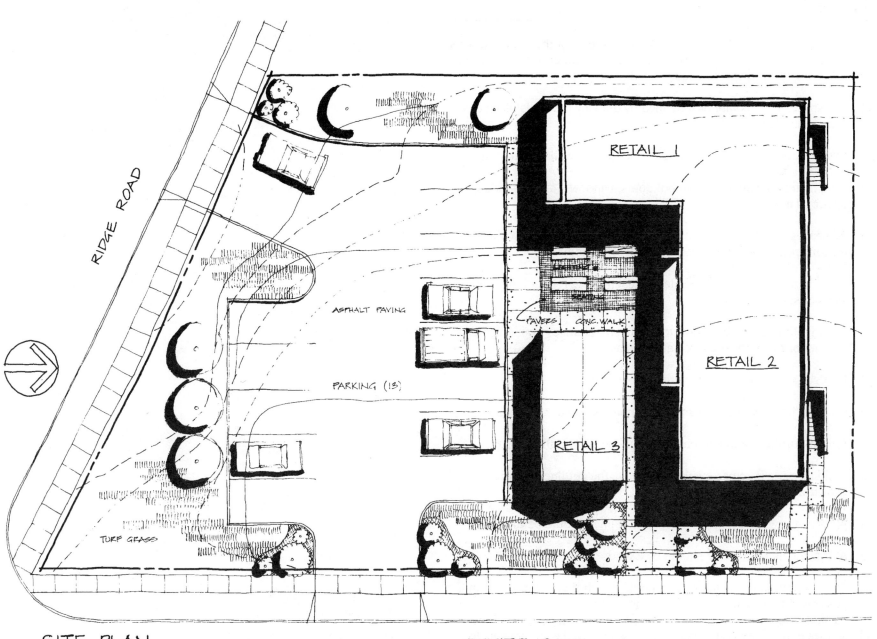

RIDGE ROAD

ASPHALT PAVING

PARKING (13)

TURF GRASS

PAVERS

CONC. WALK

RETAIL 1

RETAIL 2

RETAIL 3

SITE PLAN
1" = 10'

DEVITT ROAD

# FLOOR PLANS

For floor plans, the key is to emphasize the spaces, not the structure. This is in keeping with the overall need to show design rather than technology. The following techniques are usually used.

❑ The spaces (rooms) are best defined by drawing the walls black or heavily shaded.

❑ Exterior walls should be 8" to 12" thick, interior walls, 4" to 8" thick.

❑ Wall materials are not shown but doors, windows, and any other openings are.

❑ All spaces should be labeled with fairly large lettering. Only the drawing title should be larger, while all other notes should be smaller.

❑ Some indication of size is needed for each space (either basic dimensions or square footage).

❑ Floor materials may be drawn to emphasize changes from one area to another or to emphasize circulation areas.

❑ Landscaping should be shown if it is part of the floor plan.

❑ Changes in floor elevation should be noted with stairs and ramps.

❑ Built-in furniture and equipment should be shown and noted.

❑ Building parts (such as overhangs) above the cut line should be shown with hidden lines.

❑ A north arrow, and if applicable, section lines should be drawn.

❑ A drawing title and scale should be added.

❑ The orientation of the floor plan should match that of the site plan if possible. In the case of the project shown on the facing and previous page, the decision was made to change orientation because of the shapes of the site and the floor plan. This allowed each drawing to fit the horizontal shape of the drawing media.

❑ If there is more than one floor plan, all of them should be oriented the same regardless of building shape.

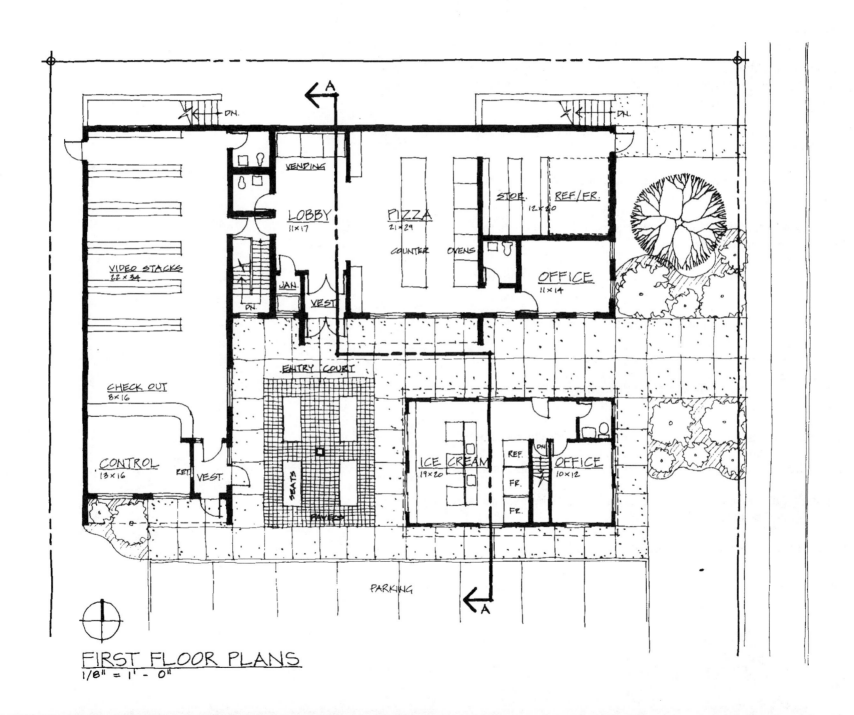

DN.

DN.

A

VENDING

LOBBY
11×17

PIZZA
21×29

STOR.
12×20

REF/FR.

VIDEO STACKS
22×34

JAN

VEST

COUNTER

OVENS

OFFICE
11×14

DN.

CHECK OUT
8×16

ENTRY COURT

CONTROL
13×16

RET.

VEST.

SEATS

PAVERS

ICE CREAM
19×20

REF.

FR.

FR.

DN.

OFFICE
10×12

PARKING

A

FIRST FLOOR PLANS
1/8" = 1' - 0"

# SECTIONS

By their nature, building sections as design drawings are very similar to floor plans. They only differ in the direction of their cut. Therefore the guidelines for drawing design sections are similar to those for walls.

❑ Walls, roof, and floors shown in section are usually filled black or heavily shaded.

❑ Wall thicknesses, including foundations, should match those in plans.

❑ The section should be taken to show important openings, whether they be in the wall, roof, or floors.

❑ Interior design features beyond the cut should be shown with thinner lines.

❑ Various floor levels and important elevations should be indicated.

❑ Interior spaces should be titled as in the plans.

❑ Miscellaneous important features should be noted with smaller lettering.

❑ People and any other objects appropriate to the project should be drawn for realism. Keep in mind, however, that we do not want to distract from the design or hinder the client's understanding of it.

❑ Sun angles, air movement, or any other such feature best shown in section should be illustrated as needed.

❑ A title and scale should be included.

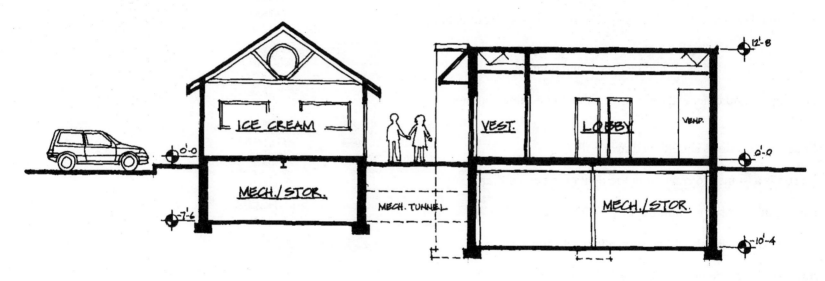

ICE CREAM

MECH./STOR.

MECH. TUNNEL

VEST.

LOBBY

VEND.

MECH./STOR.

0'-0

-7'-6

12'-8

0'-0

-10'-4

## SECTION A-A
1/8" = 1' - 0"

152 | ELEVATIONS

# ELEVATIONS

Elevations are akin to site plans. They show finish materials and the relationship of the project to the site. Many of the guidelines for drawing elevations are similar to those governing site plans.

❑ At a minimum, the main elevation and one side elevation should be drawn.

❑ The ground line should be drawn thick, with footings and foundations indicated below.

❑ Exterior finish materials should be drawn and noted.

❑ Landscaping and other exterior features should be shown but not noted.

❑ Shades and shadows should be drawn.

❑ The sun is assumed to come from the right or left at a 45° angle.

❑ Use line thicknesses to show depth.

❑ People and vehicles should be shown for scale and a sense of reality.

❑ Drawings should be titled and scaled. The scale should be the same as the plan.

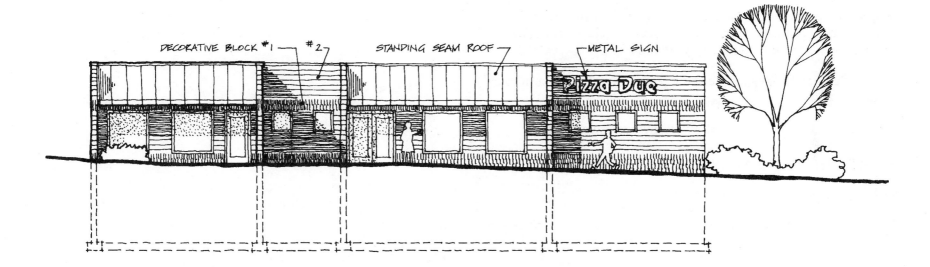

DECORATIVE BLOCK #1    #2    STANDING SEAM ROOF    METAL SIGN

Pizza Due

<u>SOUTH ELEVATION</u>
1/8" = 1' - 0"

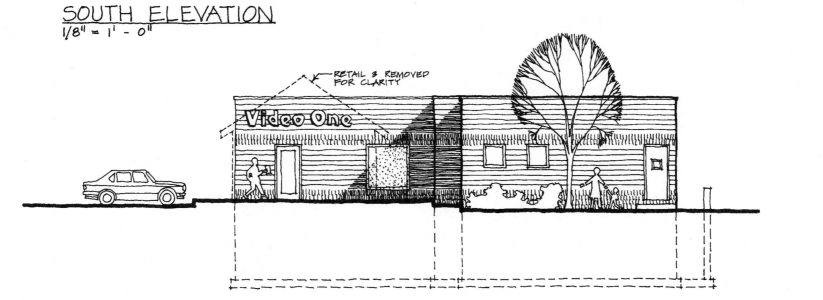

RETAIL 3 REMOVED
FOR CLARITY

Video One

<u>EAST ELEVATION</u>
1/8" = 1' - 0"

# PICTORIAL VIEWS

Perspectives, plan obliques, and less frequently, isometrics are the most commonly used pictorial views. One view should include the front. Techniques are similar to those for elevations.

❏  Materials, landscaping, and exterior features should be shown as on the elevations.

❏  Materials should not be noted unless there are no elevations.

❏  People, vehicles, etc., should be shown for realism.

❏  Shades and shadows may be shown.

❏  No scales should be indicated, but titles giving the viewpoint may be used.

VIEW FROM SOUTHWEST

# LAYOUT

Unifying all the design drawings into a package is a final but important characteristic of design drawings. One must keep in mind that the presentation can either enhance or detract from the design itself. Care and thought should be given to put the design in the best possible light.

Nonetheless a certain amount of looseness must be used in the layout. These are design drawings, not technical drawings. It is common practice to do small scale mock ups of the layout. Laying out the drawings is in itself a small design problem. It is perfectly acceptable to combine several drawings on a sheet, or all of them if they will fit. The following guidelines are a start.

❑   Title blocks are not necessary but may be used. Most important is the project title and client's name and address, and then the A/E's name and address.

❑   An indication of the design phase and sheet numbers is helpful.

❑   Drawings can be free hand or hard line.

❑   All drawings must agree with each other. This includes dimensions, cross referencing (section marks, etc.), material indications, notes, orientation, etc.

❑   Drawings should clearly explain the design to people who may not understand drawings. They should also make the client, investors, possible tenants, and town board feel excited about the project.

❑   The layout should not be over-designed so as to compete with the project design, e.g., too many borders, extraneous art work, etc. You are presenting a building design, not a graphic design. Keep it simple and effective.

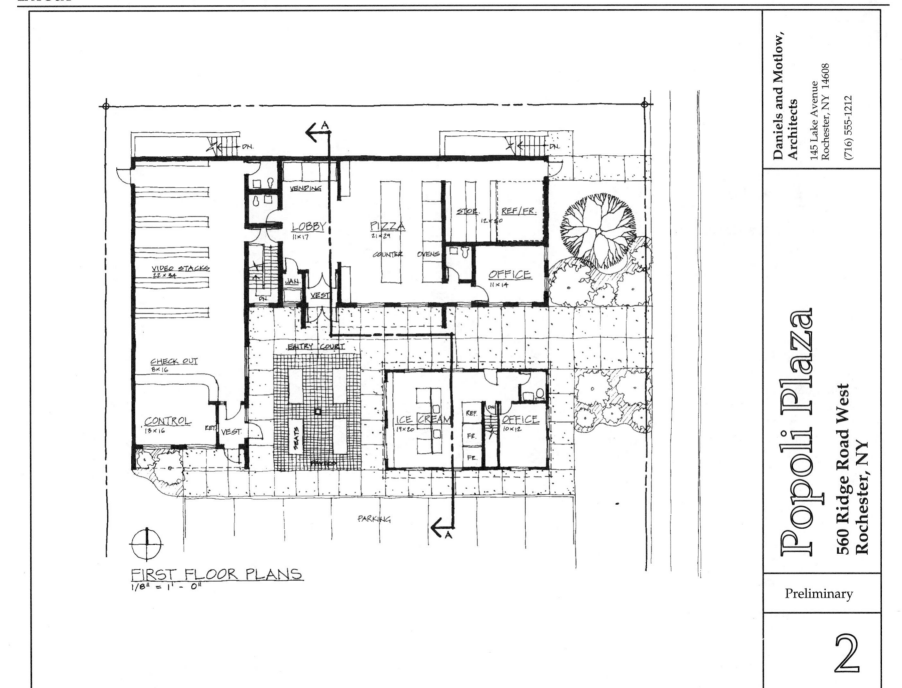

FIRST FLOOR PLANS
1/8" = 1' - 0"

**Daniels and Motlow, Architects**
145 Lake Avenue
Rochester, NY 14608

(716) 555-1212

# Popoli Plaza

**560 Ridge Road West**
**Rochester, NY**

Preliminary

2

## KEY TERMS

design drawings
preliminary drawings
presentation drawings
schematic design drawings
concept drawings

## PRACTICE EXERCISE

Draw a complete set of design drawings, including a good layout, for the project presented on this and the following pages. The drawings should include the following:

> site plan
> floor plan
> section
> the 2 main elevations
> a perspective or plan oblique

Use the site from the previous chapter. Plan to remove the garage to make room for the new building. You may locate the new building wherever you want; however the following conditions must be met:

- provide pedestian access to the front and rear doors
- include a one-way driveway from Columbus Street that provides access to the drive-in ATM and returns to Columbus Street
- provide eight to ten parking spaces off of the driveway
- add site improvements such as paving and landscaping as seem appropriate

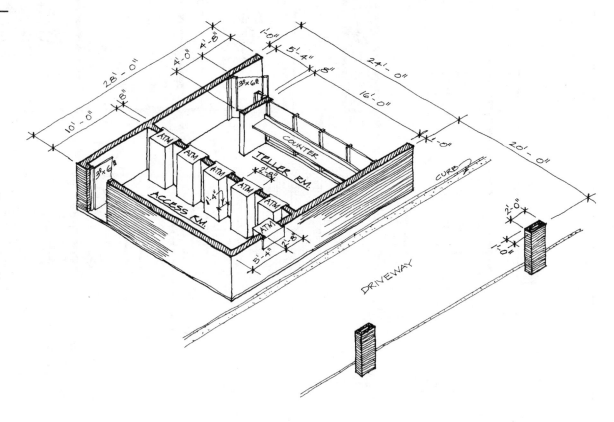

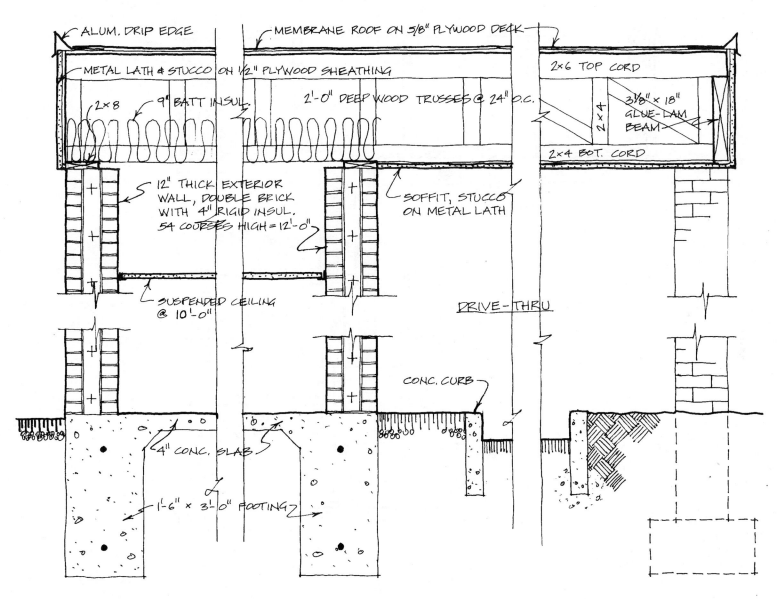

ALUM. DRIP EDGE

MEMBRANE ROOF ON 5/8" PLYWOOD DECK

METAL LATH & STUCCO ON 1/2" PLYWOOD SHEATHING

2×6 TOP CORD

2×8    9" BATT INSUL.

2'-0" DEEP WOOD TRUSSES @ 24" O.C.

2×4

3⅛" × 18" GLUE-LAM BEAM

2×4 BOT. CORD

12" THICK EXTERIOR WALL, DOUBLE BRICK WITH 4" RIGID INSUL. 54 COURSES HIGH = 12'-0"

SOFFIT, STUCCO ON METAL LATH

SUSPENDED CEILING @ 10'-0"

DRIVE-THRU

CONC. CURB

4" CONC. SLAB

1'-6" × 3'-0" FOOTING

## LONGITUDINAL SECTION

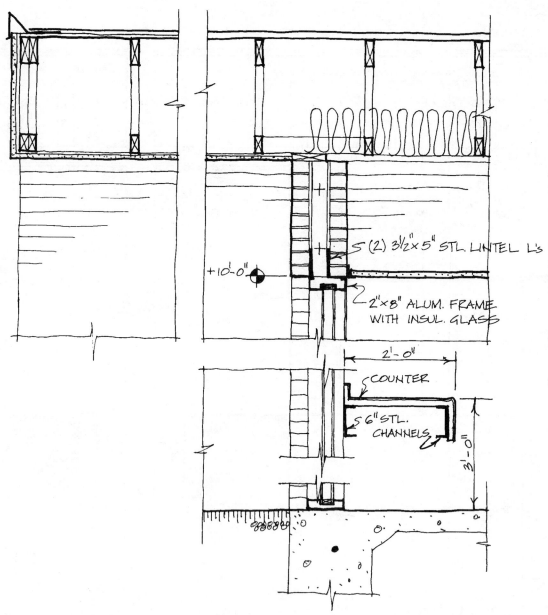

(2) 3½" x 5" STL. LINTEL L's

+10'-0"

2" x 8" ALUM. FRAME
WITH INSUL. GLASS

2'-0"

COUNTER

6" STL.
CHANNELS

3'-0"

## SECTION AT ENTRANCE

# PART III

# CONSTRUCTION DRAWINGS

# 13

# CONSTRUCTION DRAWINGS

## • OVERVIEW

### OBJECTIVES

*By the end of this chapter students should be able to:*
- Lay out a set of working drawings in a logical sequence.
- Produce a cover sheet.
- Produce a legend and key plan.
- Cross reference a set of drawings.
- Write effective, efficient notes.

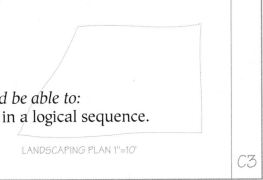

LANDSCAPING PLAN 1"=10'

C3

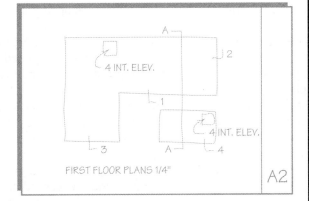

FIRST FLOOR PLANS 1/4"

A2

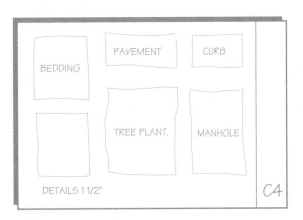

BEDDING    PAVEMENT    CURB

TREE PLANT.    MANHOLE

DETAILS 1 1/2"

C4

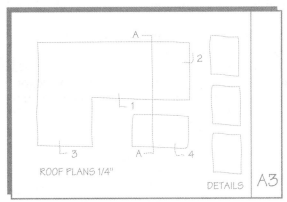

ROOF PLANS 1/4"

DETAILS

A3

The contract document phase of a construction project is the most time-consuming phase for an A/E office. Preparing the working drawings and specifications is a critical component in the design process, and it requires skill and patience. One of the most important requirements for this process is a high degree of organization.

This chapter introduces the tools used to organize a set of construction drawings. It will use the project seen in the past few chapters as an example. The same project is used in the final three chapters to illustrate the actual execution of the drawings.

# SET ORGANIZATION

The drawings and specifications must work together to describe the project completely. They must explain the project so that the contractors and subcontractors that use the drawings can understand them. In a sense, the drawings must speak the language of carpenters, masons, electricians, etc.

The drawings should be arranged by trades, which reflects to a certain degree the order in which the project is completed. Each trade prefers to have its information presented in a manner that is specifically useful to them.

Secondly, but perhaps more importantly, drawings must be cross referenced. **Cross referencing** is the system of relating one drawing to another. By using a system of reference symbols, a set of drawings can be read like a book.

For example, the drafter places section lines on a plan that tell the person reading the plan in which direction it is viewed and where to find the related section drawing. A similar symbol on the section would lead the reader to an enlarged detail to show very precise construction information.

Using such a system is like providing road signs for readers of the drawings. It helps them navigate about the drawings to get the information they need. Without cross referencing it would be left to fate as to whether the readers found what they were looking for.

Following is an outline for the order in which drawings should be arranged in a set.

Cover Sheet
    Index, Legend, Materials, Referencing
Civil/Site
    Site Improvement Plan
    Grading Plan
    Landscaping Plan
Architectural
    Floor Plans
    Roof Plan
    Schedules
    Building Sections
    Wall Sections
    Exterior Elevations
    Enlarged Plans and Interior Elevations
    Reflected Ceiling Plans
    Details
Structural
    Foundation Plan
    Foundation Sections, Elevations, Details
    Framing Plans and Schedules
    Framing Sections/Elevations
    Framing Details
Mechanical
    Site Plan
    Floor Plan and Schedules
    Details
Electrical
    Site Plan
    Floor Plan and Schedules
    Details
Plumbing
    Site Plan
    Floor Plan and Schedules
    Details

## Cover Sheet

Unlike other sheets in a set of working drawings, cover sheets often do not have a title block. The project title and the A/E office that designed it are the first items to be included.

Cover sheet information should help the reader find and then understand everything in the set. Typically cover sheets include the following as illustrated below:
- index or list of drawings
- index or list of material symbols
- index or list of reference symbols

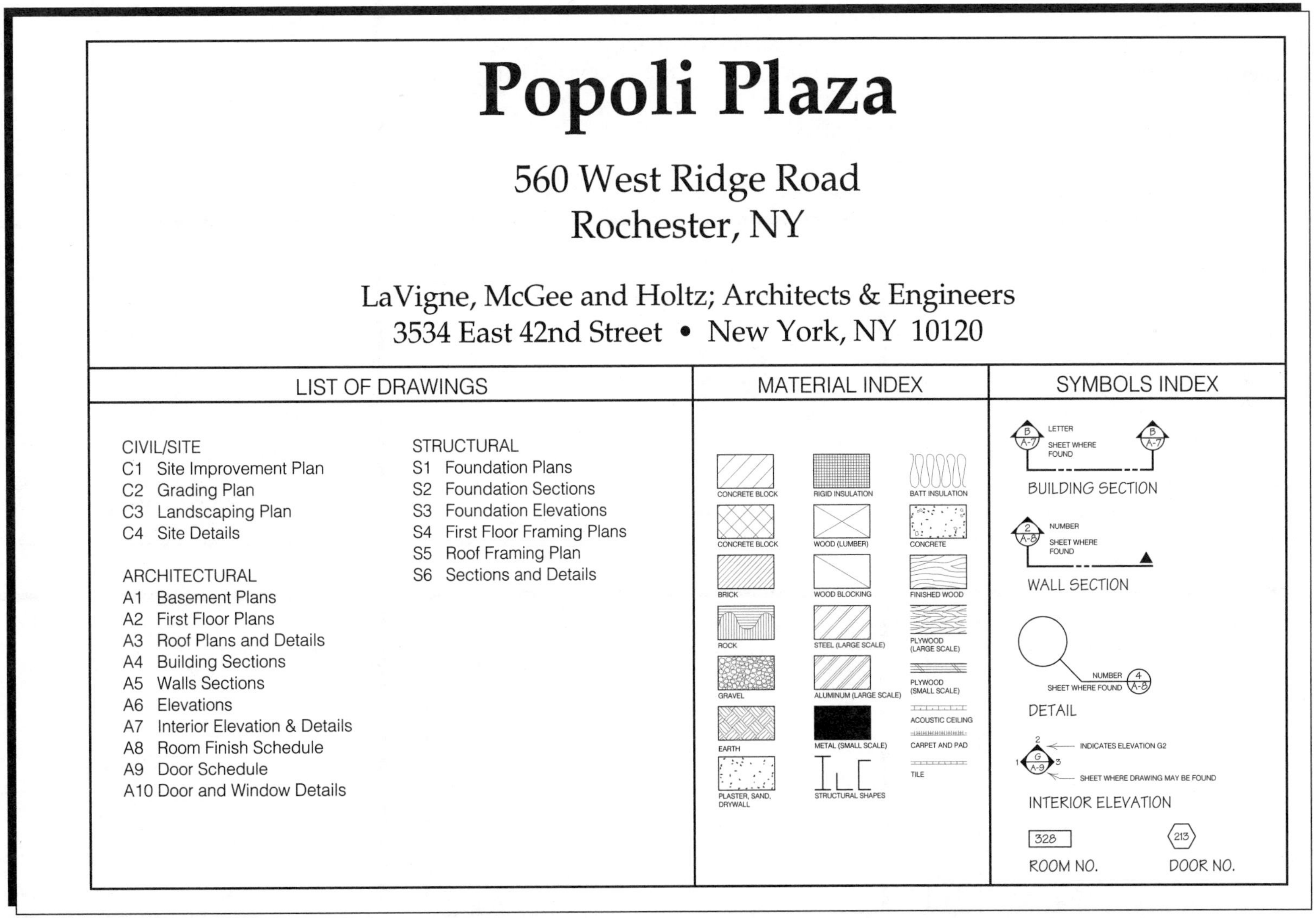

# Popoli Plaza

### 560 West Ridge Road
### Rochester, NY

### LaVigne, McGee and Holtz; Architects & Engineers
### 3534 East 42nd Street • New York, NY 10120

| LIST OF DRAWINGS | MATERIAL INDEX | SYMBOLS INDEX |
|---|---|---|

**LIST OF DRAWINGS**

CIVIL/SITE
C1 Site Improvement Plan
C2 Grading Plan
C3 Landscaping Plan
C4 Site Details

ARCHITECTURAL
A1 Basement Plans
A2 First Floor Plans
A3 Roof Plans and Details
A4 Building Sections
A5 Walls Sections
A6 Elevations
A7 Interior Elevation & Details
A8 Room Finish Schedule
A9 Door Schedule
A10 Door and Window Details

STRUCTURAL
S1 Foundation Plans
S2 Foundation Sections
S3 Foundation Elevations
S4 First Floor Framing Plans
S5 Roof Framing Plan
S6 Sections and Details

**MATERIAL INDEX**

CONCRETE BLOCK — RIGID INSULATION — BATT INSULATION
CONCRETE BLOCK — WOOD (LUMBER) — CONCRETE
BRICK — WOOD BLOCKING — FINISHED WOOD
ROCK — STEEL (LARGE SCALE) — PLYWOOD (LARGE SCALE)
GRAVEL — ALUMINUM (LARGE SCALE) — PLYWOOD (SMALL SCALE)
EARTH — METAL (SMALL SCALE) — ACOUSTIC CEILING / CARPET AND PAD
PLASTER, SAND, DRYWALL — STRUCTURAL SHAPES — TILE

**SYMBOLS INDEX**

LETTER / SHEET WHERE FOUND
BUILDING SECTION

NUMBER / SHEET WHERE FOUND
WALL SECTION

NUMBER / SHEET WHERE FOUND
DETAIL

INDICATES ELEVATION G2 / SHEET WHERE DRAWING MAY BE FOUND
INTERIOR ELEVATION

328 ROOM NO.     213 DOOR NO.

## Layout

Often the job captain on a project will lay out an entire set of working drawings as small scale, free-hand thumbnail sketches before beginning the actually drawings. This helps organize all the drafters that may be involved as the project develops. It also promotes a comprehensive, planned set of drawings.

The illustrations that follow show this technique. They also provide an opportunity to show the example project in its entirety. As one studies the chapters that follow, a return to these few pages will help keep the work in perspective.

A final purpose of this small layout is to show how the cross referencing system works. The specific symbols follow in the next section but can be seen as a whole here.

Note here that drawings are numbered according to discipline: "C" for civil/site, "A" for architectural, "S" for structural, etc. Mechanical, electrical, and plumbing drawings are not shown since they will not be covered in this text.

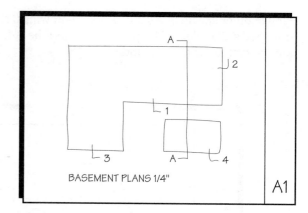

BASEMENT PLANS 1/4"    A1

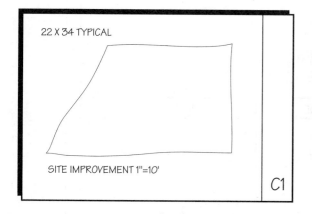

22 X 34 TYPICAL

SITE IMPROVEMENT 1"=10'    C1

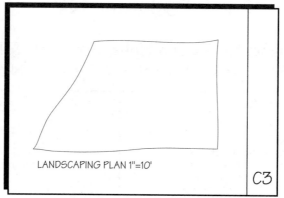

LANDSCAPING PLAN 1"=10'    C3

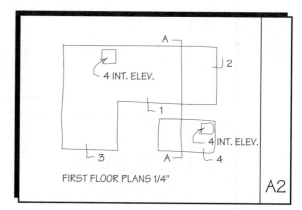

FIRST FLOOR PLANS 1/4"    A2

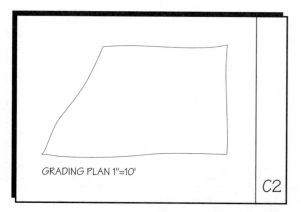

GRADING PLAN 1"=10'    C2

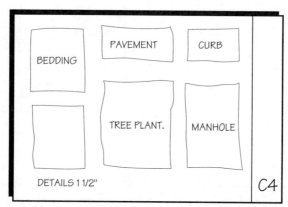

BEDDING    PAVEMENT    CURB

TREE PLANT.    MANHOLE

DETAILS 1 1/2"    C4

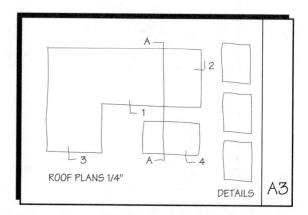

ROOF PLANS 1/4"    DETAILS    A3

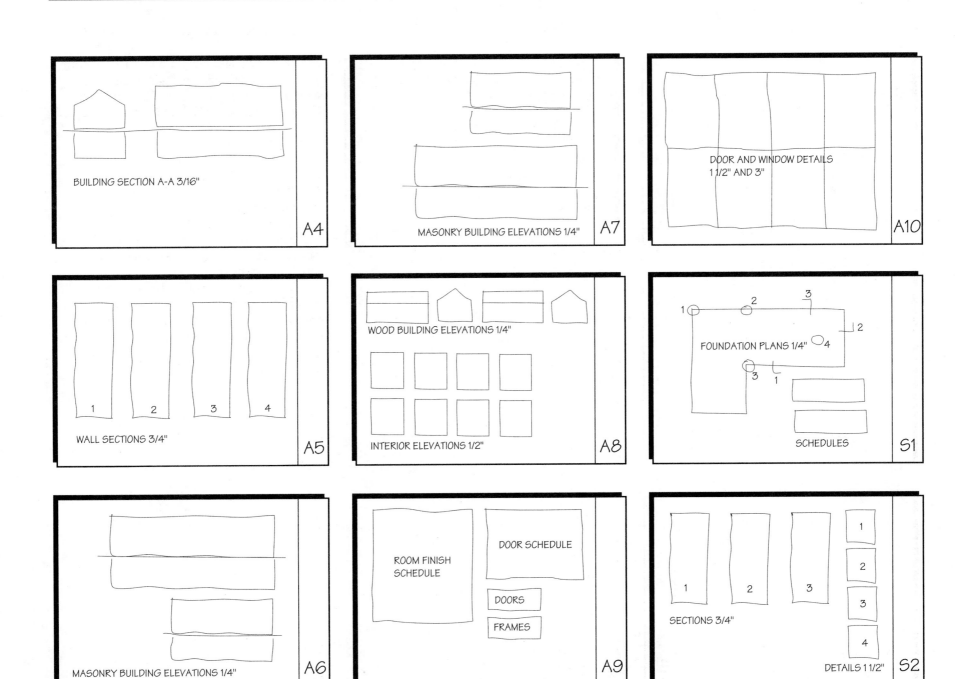

BUILDING SECTION A-A 3/16"  A4

MASONRY BUILDING ELEVATIONS 1/4"  A7

DOOR AND WINDOW DETAILS
1 1/2" AND 3"  A10

WALL SECTIONS 3/4"  A5

WOOD BUILDING ELEVATIONS 1/4"

INTERIOR ELEVATIONS 1/2"  A8

FOUNDATION PLANS 1/4"

SCHEDULES  S1

MASONRY BUILDING ELEVATIONS 1/4"  A6

ROOM FINISH
SCHEDULE

DOOR SCHEDULE

DOORS

FRAMES  A9

SECTIONS 3/4"

DETAILS 1 1/2"  S2

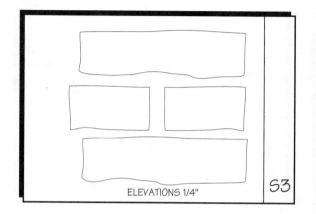

ELEVATIONS 1/4"

S3

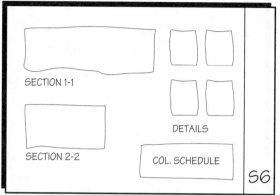

SECTION 1-1

SECTION 2-2

DETAILS

COL. SCHEDULE

S6

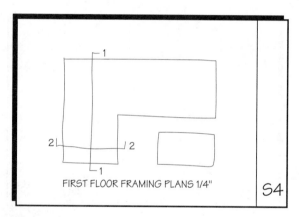

FIRST FLOOR FRAMING PLANS 1/4"

S4

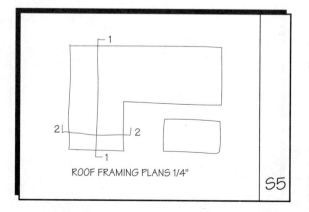

ROOF FRAMING PLANS 1/4"

S5

## Cross Referencing

The next few pages illustrate in schedule form most of the cross referencing and labeling symbols used on a set of working drawings. This particular set was developed and adopted by an actual A/E office. Every drafter in that office is expected to know and use these symbols.

Since there is not a set of universally accepted symbols in the AEC industry, it is common for offices to develop their own. Standards tend to vary by region. However if one were to collect samples of the ones in use today, there would not be a great deal of variation among them.

What is important is to be consistent within a set of drawings and surely within an office setting. You may use the symbols presented here as a starting point.

| CROSS REFERENCING, SYMBOL TYPE | SIZE | EXAMPLE |
|---|---|---|
| **DIMENSIONS**<br><br>• MASONRY TO FACES<br>• ALL OTHERS TO CENTER LINE<br><br>• DOTS FOR COLUMN CENTER LINES<br>• ARROWS FOR NARROW DIMENSIONS<br>• SLASHES FOR ALL OTHERS | | 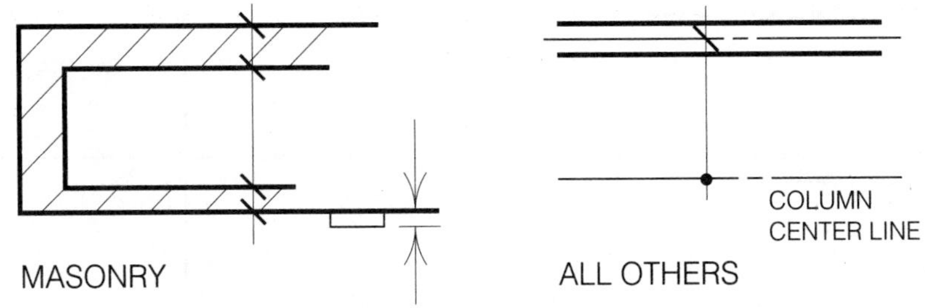 MASONRY　　　　ALL OTHERS<br>COLUMN CENTER LINE |
| **BUILDING CROSS SECTION**<br><br>• LETTERS: A-A, B-B, ETC.<br>• ON PLANS ONLY | 3/8" DIA. | 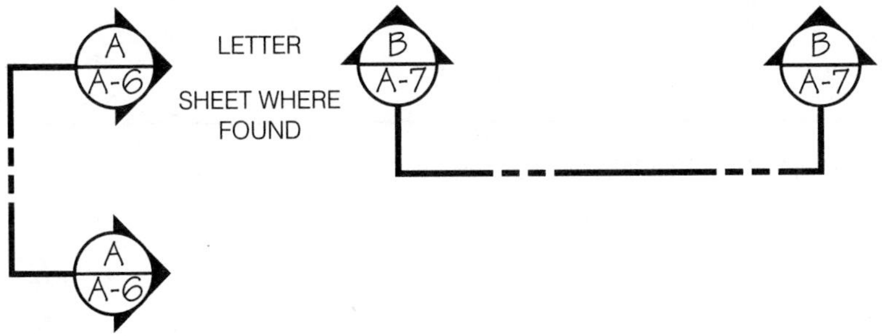 LETTER<br>SHEET WHERE FOUND |
| **DETAIL SECTION**<br><br>• NUMBERS: 1, 2, 3, ETC.<br>• ON PLANS AND ELEVATIONS | 3/8" DIA. | 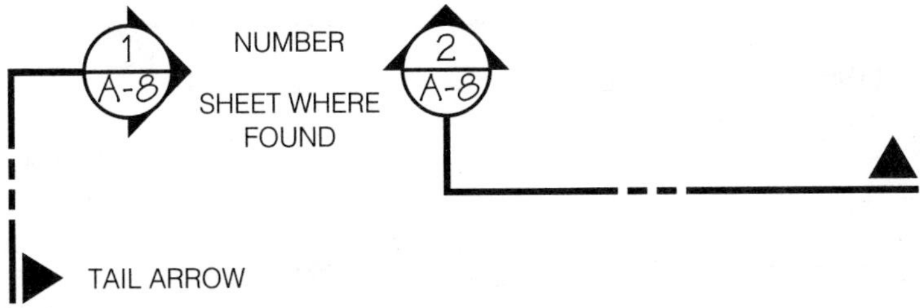 NUMBER<br>SHEET WHERE FOUND<br>TAIL ARROW |
| **DETAIL ENLARGEMENT**<br><br>• NUMBERS 1, 2, 3, ETC.<br>• ON PLANS, ELEVATIONS, AND SECTIONS | AS REQUIRED<br><br>3/8" DIA. | 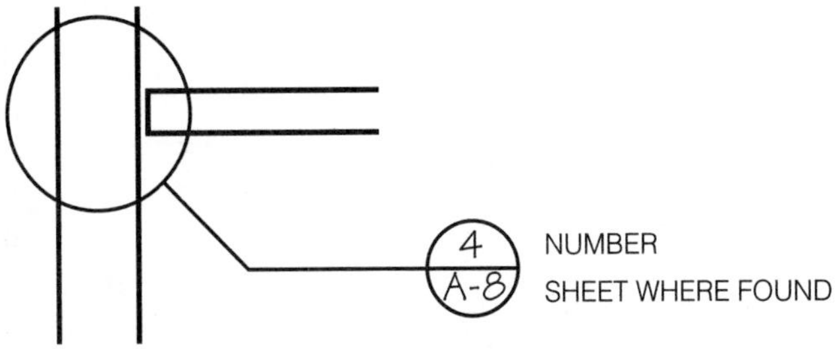 NUMBER<br>SHEET WHERE FOUND |

| CROSS REFERENCING, SYMBOL TYPE | SIZE | EXAMPLE |
|---|---|---|
| COLUMN LINE | 3/8" DIA. | 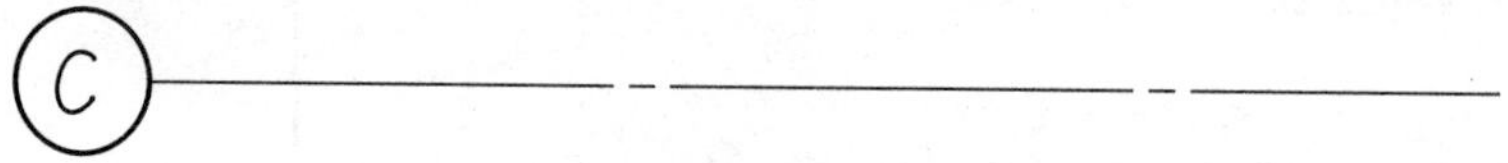 |
| DETAIL TITLE<br><br>• NUMBERS: 1, 2, 3, ETC. | 1/2" DIA. |  **BASE DETAIL**<br>1 1/2" = 1' - 0"<br>NUMBER<br>5<br>A-3<br>LOCATION OF PARENT DRAWING |
| INTERIOR ELEVATIONS<br><br>• ON PLANS ONLY<br>• LETTERS AND NUMBERS, E.G. ELEVATION G1, G2, G3 | 3/8" DIA. | 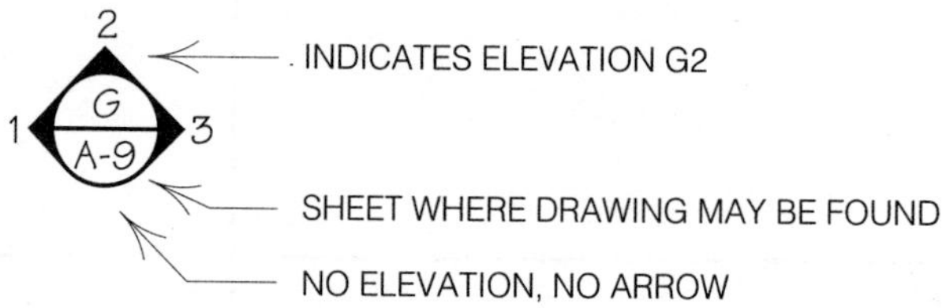 2 ← INDICATES ELEVATION G2<br>1 G 3<br>A-9<br>SHEET WHERE DRAWING MAY BE FOUND<br>NO ELEVATION, NO ARROW |
| ROOM NUMBER AND TITLE<br><br>• ON PLANS ONLY | 1/4" x REQUIREMENT |  **OFFICE**<br>328 |
| DOOR NUMBER<br><br>• ON PLANS, ONLY | 3/8" DIA. | 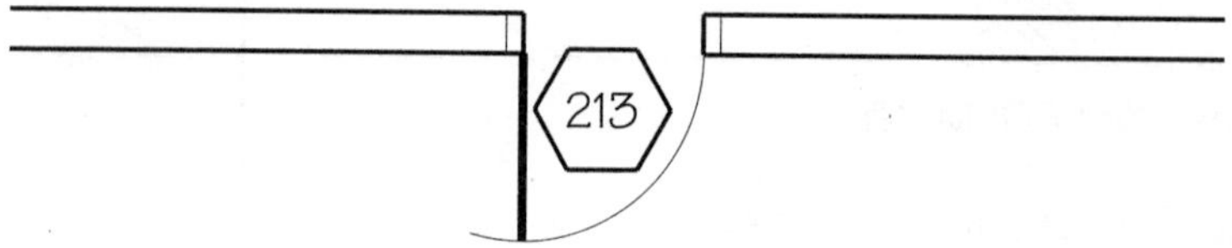 213 |
| ELEVATION MARK<br><br>• ON ELEVATIONS<br>• ON SECTIONS AND DETAILS AS REQUIRED | 1/4" DIA. | 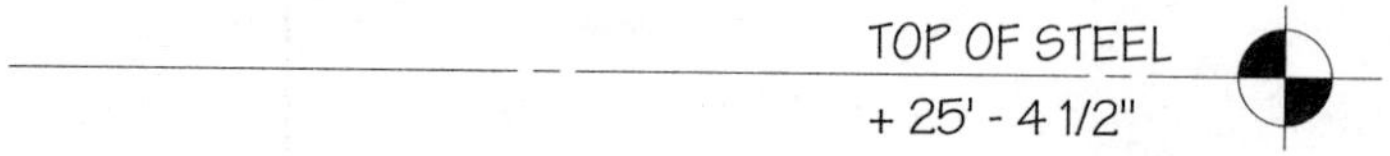 TOP OF STEEL<br>+ 25' - 4 1/2" |

| CROSS REFERENCING, SYMBOL TYPE | SIZE | EXAMPLE |
|---|---|---|
| **NORTH ARROW**<br><br>• ON PLANS AND SITE PLANS ONLY<br>• ADD TRUE NORTH LINE WHEN ORIENTATION IS SKEWED | 3/4" TO 1" DIA. | 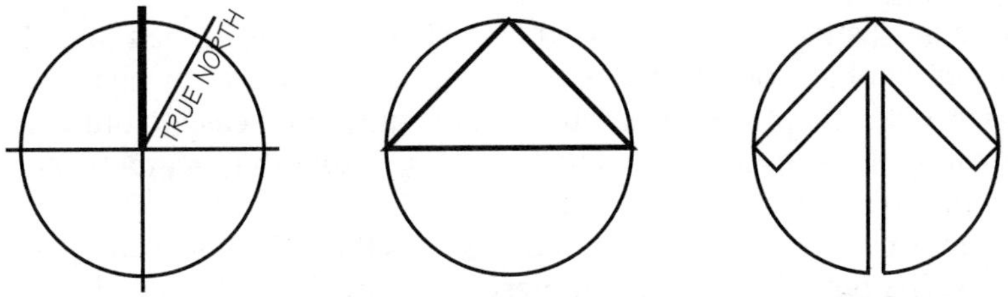 TRUE NORTH |
| **LETTERING**<br><br>• GENERAL NOTES<br>  ARROWS TO FACE<br>  DOTS TO AREAS | 3/32" | |
| • DRAWING SUBTITLES OR ROOM NAMES, ETC. | 3/16" | |
| • DIMENSIONS OR SPECIAL NOTES | 1/8" | |
| • DRAWING TITLE | 1/4" | |
| • SCALE | 1/8" | |
| • SHEET NUMBER, MACHINE MADE OR HAND OUTLINE | 1/2" TO 1" | |

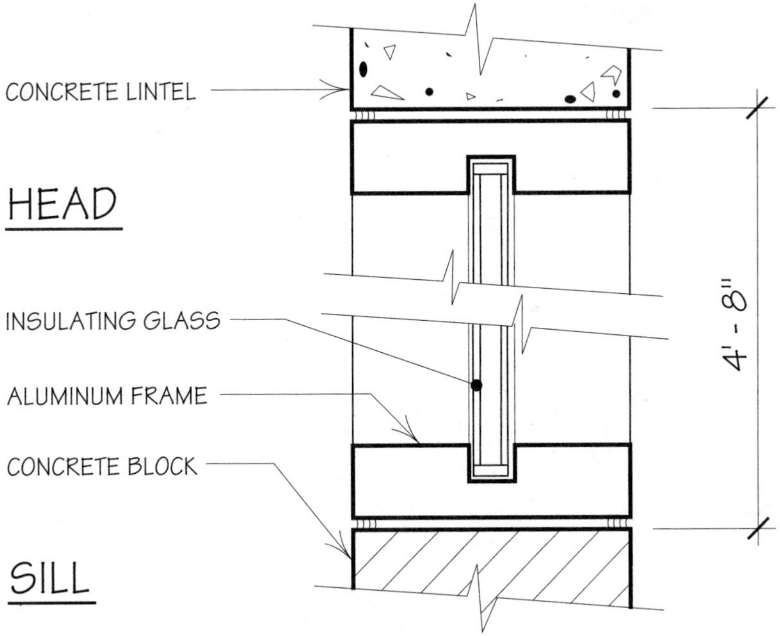

CONCRETE LINTEL

HEAD

INSULATING GLASS

ALUMINUM FRAME

CONCRETE BLOCK

SILL

4' - 8"

# WINDOW DETAILS
3" = 1' - 0"

A-1

# LEGENDS

A *legend* is a listing of symbols used on a sheet along with their meaning. Whenever an object or material is used extensively on a drawing it saves time to give it a simple representation that is quick and easy to draw. The object can be identified in the legend one time rather than many times throughout the drawing.

Legends can identify more than objects, as the following examples illustrate. The point is that every representational element drawn by a drafter needs to be defined. This is done by defining materials on the cover and by including legends throughout the remainder of the set.

## LEGEND

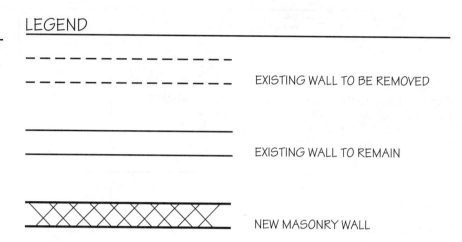

EXISTING WALL TO BE REMOVED

EXISTING WALL TO REMAIN

NEW MASONRY WALL

# KEY PLANS

A *key plan* is a small scale outline of the building that identifies the portion of the floor plan that appears on that sheet. They are utilized for large buildings that cannot be drawn onto a single sheet but must be broken into segments and drawn over several sheets.

A key plan is usually incorporated into the title block but can be drawn anywhere on the sheet as space allows. It should be at the same location on each sheet however.

A key plan may also be used when a portion of any plan is enlarged to show detail. As a cross referencing measure we need to show where on the overall plan this one area came from.

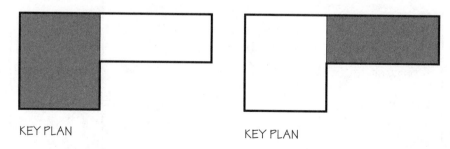

KEY PLAN                                    KEY PLAN

*If a floor plan is too big to fit onto one sheet it must be split. The key plan on the left would be found on the sheet containing the left half of the building, and the key plan on the right would be found on the sheet containing the right half.*

## LEGEND

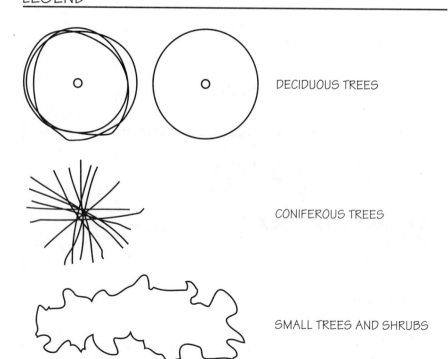

DECIDUOUS TREES

CONIFEROUS TREES

SMALL TREES AND SHRUBS

# NOTES

There are three types of notes used on working drawings:
- general notes
- material identification notes
- instructions

## General Notes

General notes are perhaps the easiest to comprehend and are very similar in concept to legends. For many drawings in a set there is often information that applies to the entire sheet. A single note that applies to the whole drawing is more efficient than restating the information many times.

General notes are grouped together, usually numbered, and given the heading "General Notes" or just "Notes." They are located wherever there is open space but should be prominent enough so that they cannot be missed.

Following is an example for a drawing depicting a wood building. Since the building has many different types of wood components, some notes apply to all parts, others to a sub-group of components.

GENERAL NOTES
1.  ALL LUMBER TO BE HEM-FIR 900 PSI STRESS GRADE
2.  ALL PLYWOOD TO BE EXTERIOR GRADE
3.  CONTRACTOR TO PROTECT ALL WOOD PRODUCTS ON SITE FROM THE WEATHER UNTIL INSTALLED
4.  ROOF SHEATHING TO BE JOINED AT EDGES WITH CLIPS
5.  JOIST HANGERS MAY BE USED IN PLACE OF TRADITIONAL NAILING AT CONTRACTOR'S OPTION

## Material Identification Notes

At the opposite end of the spectrum from general notes are material indication notes. Every individual component of an entire structure must be labeled.

One might remember from an earlier chapter that the purpose of working drawings is to show "what" and "where." Material indication notes that identify objects essentially take care of the "what."

As the following chapters will show, these notes comprise the greatest amount of lettering on working drawings. It should be clear that it is important for drafters to write their notes well.

Material identification notes may contain up to four fields of information about an object:
- its size or other descriptor
- the material it is made of
- its name
- its frequency

All notes need to include the **name** of the object. The simplest note might only need the name. This would be the norm for pre-assembled objects or units made of several materials. Examples include:

HEATING UNIT
PRE-FABRICATED FIREPLACE
FLAG POLE

The note identifies these manufactured items, and the dimensions locate them. The specifications will explain the "who" and the "how"—that is, who the manufacturers are, the model numbers, and how the contractor is to install them.

For components that are assembled on site, we need to add the **material** to our note. For example, the note "BLOCK" is not descriptive enough. We need to say what kind of block we want. Most material identifications fall into this category, such as:

CONCRETE BLOCK
PLYWOOD SHEATHING

STEEL COLUMN

This type of note works well until a project is large enough to have several types of one material. In this case another descriptor is needed to distinguish one from another. The most common example of this is when materials come in many **sizes**.

Suppose 8" and 12" concrete blocks are being used on a project. The drafter needs to show where each thickness is used by noting it as such. Notes in this situation might now read:

> 8" CONCRETE BLOCK
> 3/8" PLYWOOD SHEATHING
> 3" DIAMETER STEEL COLUMN

If size does not distinguish two types, then whatever it is that makes them different needs to be stated. If we had two types of concrete block on a project and both were 8", their difference might be noted as:

> SPLIT FACE CONCRETE BLOCK
> STANDARD CONCRETE BLOCK

The final descriptor is used for objects that are not continuous. Structural members, for example, are spaced at intervals. We therefore need to note their **frequency.**

> 2 x 10 WOOD JOISTS AT 16" ON CENTER
> 1/2" STEEL ANCHOR BOLTS AT 6'-0" ON CENTER
> STEEL TRUSS REINFORCING EVERY OTHER COURSE

There is a preferred order to descriptors in a note, namely: size (or style), material, name, and frequency, as illustrated below:

| SIZE | MATERIAL | NAME | FREQUENCY |
|------|----------|------|-----------|
| #5 | STEEL | REINFORCING BAR | |
| 4" | CONCRETE | SIDEWALK | |
| 2 x 4 | WOOD | STUDS | @ 16" O.C. |
| | VINYL | FLOORING | |
| | | LIGHT FIXTURE | |
| 2" RED | CERAMIC | TILE | |
| | | CARPET | |
| 1/2" | STEEL | BOLTS | @ CORNERS |

## Instructions

The final type of note is that which gives specific instructions to the contractor, usually pertaining to installation. This type of note is usually kept to a minimum because specifications are the place to explain "how." Nonetheless, there are situations where showing the exact location of an instruction is very important.

The following examples illustrate where such instructions might occur and how they should be written. Notice that the notes are not written as complete sentences but rather as short, direct instructions.

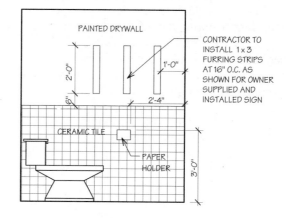

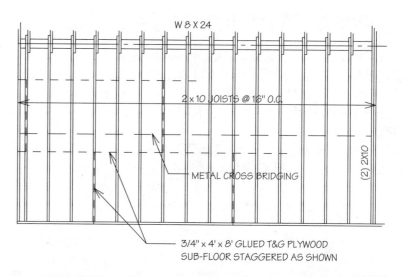

# KEY TERMS

cross reference
legend
key plan

# PRACTICE EXERCISE

For the practice exercise project in Chapter 12 (Design Drawings), complete the following:

- lay out with thumbnail sketches the working drawings that might be needed for construction; select a sheet size and drawing scales
- indicate on your layout the cross referencing between drawings
- design and draw a cover sheet

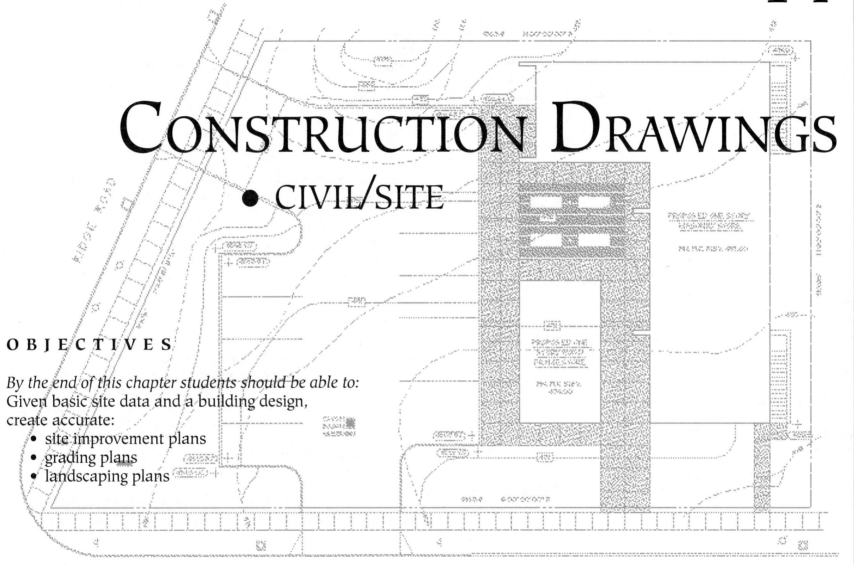

# 14

# CONSTRUCTION DRAWINGS
- CIVIL/SITE

**OBJECTIVES**

*By the end of this chapter students should be able to:*
Given basic site data and a building design,
create accurate:
- site improvement plans
- grading plans
- landscaping plans

Civil/site drawings generally consist of three types: site plans, *profiles* (full sections through the site), and details. Profiles are used almost exclusively for pure engineering projects such as highways and railroads. Site plans and details are the only civil/site drawings generally needed for building construction projects.

Work on the site involves three basic areas: site improvements, grading, and landscaping. Sometimes a different subcontractor is responsible for each, so it is often prudent to make a plan for each, namely:

- site improvement plan
- grading plan
- landscaping plan

For smaller projects it is quite possible to combine all three of these drawings onto one site plan. Or if two drawings can be combined without confusion, this is also acceptable. Once again the drafter needs to balance efficient use of drafting materials with a clear presentation of the intended information.

In a set of working drawings, the site drawings usually come first. It must be noted here that site utilities are a very important part of the site design. Utilities also have different subcontractors for each discipline: one for electrical work, one for plumbing work, and one for mechanical work. On a large project there may be a separate utility site plan for each of these disciplines. These plans prepared by the mechanical and electrical engineers for the project and site plans are placed with their respective disciplines in a set.

Let us now turn our attention to the site drawings involving civil engineering and landscape architecture.

# SITE IMPROVEMENT PLAN

Site improvements refer to all the manmade features of the design that are part of the project except for buildings. Thus, items such as parking lots, piping for drainage, paved areas, sidewalks, retaining walls, fencing, and the like would all be considered site improvements.

The objective of this drawing is to show each element, dimension its location, and name the materials used. For this, as well as other site drawings, it is necessary to clearly indicate which items are new, which are existing to be removed, and which are existing to remain. Careful wording of the notes and use of symbols will accomplish this.

## Property Lines

The first step in preparing a site plan is to outline the property with a property line. The line thickness is usually medium; however, many drafters prefer drawing it thick.

The property should be oriented so that north is to the top of the sheet (preferred) or to the right. It is also common practice to place the drawing towards one of the corners of the sheet, thereby leaving space for notes and details.

An alternate and commonly used convention is to place the entrance side of the project to the bottom of the sheet regardless of compass direction.

Site plans are almost always drawn at one of the engineering scales, i.e., 1" = 10', 20', 30', 40', 50', or 60'. Like any other construction drawing, the scale should be large enough to show the information clearly but small enough to fit onto a standard sheet size.

The bearing and length of each property line should be taken from the survey map from which the site plan is being prepared. It is often helpful to place the north arrow on the drawing at this time to assist in drawing the bearings.

## Building Footprint

The next step of the drawing is to outline the building. What is important here is to draw what is called the building *footprint*, rather than what the top of the building looks like. A footprint is the outline of a building where it contacts the ground. The concept springs from a person's footprint, which is a different shape than the outline of a person's body from above.

Because generally there is some sort of overhang on a building, the roof plan is often larger than the footprint. More importantly, there may be site features under an overhang that could not be shown if the roof plan were drawn.

The building footprint should be drawn with a thick object line since it represents a significant projection from the surface of the property.

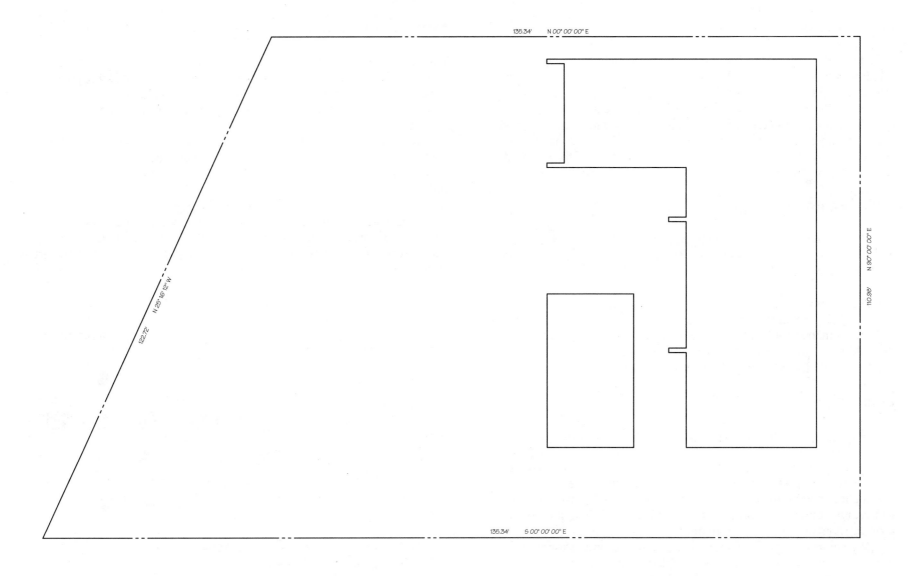

135.34'   N 00° 00' 00" E

122.72'   N 25° 16' 12" W

110.98'   N 90° 00' 00" E

135.34'   S 00° 00' 00" E

## Paved Areas

Paved areas may be added after the property is delineated and the building is blocked out. Roadways, driveways, parking lots, and sidewalks often take up most of the remaining space. And more frequently today patios and decks are also added as site improvements.

Paved areas should be drawn as object lines of medium weight since they project very little from the surface. Joints or painted lines on the surfaces should be drawn as thin or very thin lines. If curbs are used, they should also be drawn to scale.

## Other Features

All other manmade features may be added once the paved areas are defined. These features include such things as fixed furniture, fencing, art work, retaining walls, and any other manmade object placed on the site, except utilities.

These items should also be drawn as object lines of medium weight, but could also appear as hidden lines if they are below ground or otherwise not directly visible. Natural elements such as lawns and plants are drawn on the landscaping plan, and utilities are drawn on the mechanical, electrical, and plumbing site plans.

## Dimensions

Site improvement plans can usually be dimensioned using the basic conventions presented in Part I of the text. The same general rule applies: Use as few extension and dimension lines as possible to avoid clutter.

The most important dimensions are those that fix the building at a specific location on the site. To do this, two dimensions will usually suffice. One dimension should establish the setback from the front or rear property line, and the other from one of the side lot lines.

The complexity of the dimensioning system is proportional to the number of features that are on the site plan. Road widths and radii of curves need to be dimensioned. All parking spaces need to be dimensioned; however a "typical" dimension for one space is usually adequate.

Other paved areas like walkways, patios, decks, etc., need to be dimensioned. To check whether you have provided adequate dimensions on the plans, put yourself in the place of the contractor and see if there is enough data from which to build.

## Material Indication

The next step is to pattern all of the improvements. The two most common material indications are used in our example on the facing page. Asphalt paving is generally shown with no patterning. The pattern for concrete is simply a dot pattern. The appendix shows additional material indications.

A convention for patterning is to render only new materials and leave existing ones blank. This helps determine what is new and what is existing. The notes that accompany the object will also make the distinction.

One can see in the example on the facing page that the existing sidewalks that run parallel to the streets are blank. The new concrete walks are patterned. If an existing feature needs to be removed, it is shown with hidden lines and noted accordingly. Typical conventions are shown below.

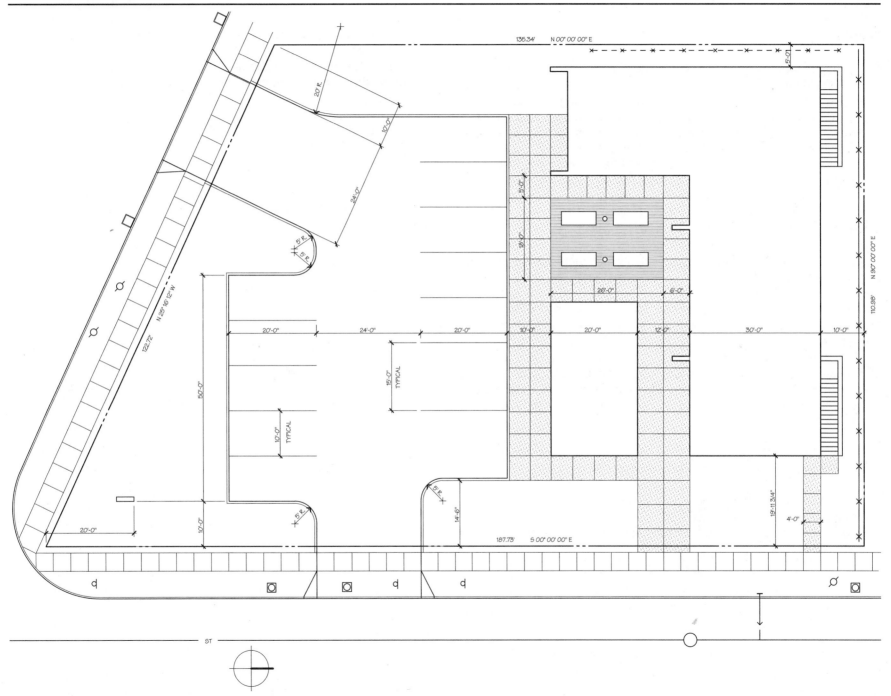

## Notes

Everything that has been drawn on the site plan must now be noted. Most notes identify objects, state their material, and say whether the objects are proposed, existing to be removed, or existing to remain.

For existing materials that are to remain there are two common ways of noting them. As shown in the example on the facing page, the existing sidewalks are noted as EXISTING CONCRETE SIDEWALK. It would also be acceptable to note them as EXISTING CONCRETE SIDEWALK TO REMAIN.

Existing materials to be removed are handled a little differently. Whether or not the object has been drawn with the broken line convention, it is common to state that the object is to be removed. In our example the wire fence has been noted in such a way:
EXISTING METAL FENCE TO BE REMOVED.

When the note identifies an improvement then there are two conventions that are followed and either one is acceptable. It is assumed that all materials are proposed unless otherwise noted. Some offices insert the term PROPOSED at the beginning of the note; others leave it off. Both PROPOSED CONCRETE SIDEWALK and CONCRETE SIDEWALK are acceptable notes for a new sidewalk.

Structures on the site should be noted with large underlined lettering (1/8" to 3/16"). The first floor elevation is also usually noted under the title at the next smaller size lettering. The title itself is worded to include the type of construction, the number of floors, and the type of structure. Examples would be:

ONE STORY FRAME HOME
TWO STORY MASONRY APARTMENT BUILDING

Identification of streets should also be noted in large lettering. This lettering is usually not underlined but is often spread out to run along most of the street it names. The example on the facing page illustrates this convention.

The remaining notes identify surfaces and minor objects. Lettering should be small (1/8" or 3/32") and should include the material where applicable. Leaders and terminators are frequently used to clearly indicate where the object or surface is located. Examples include:

WOOD BENCH
STEEL BOLLARD
SAPLING FENCE
METAL SCULPTURE

The final lettering is the drawing title and scale, which follow previously established guidelines. A north arrow should be added if it was not added earlier.

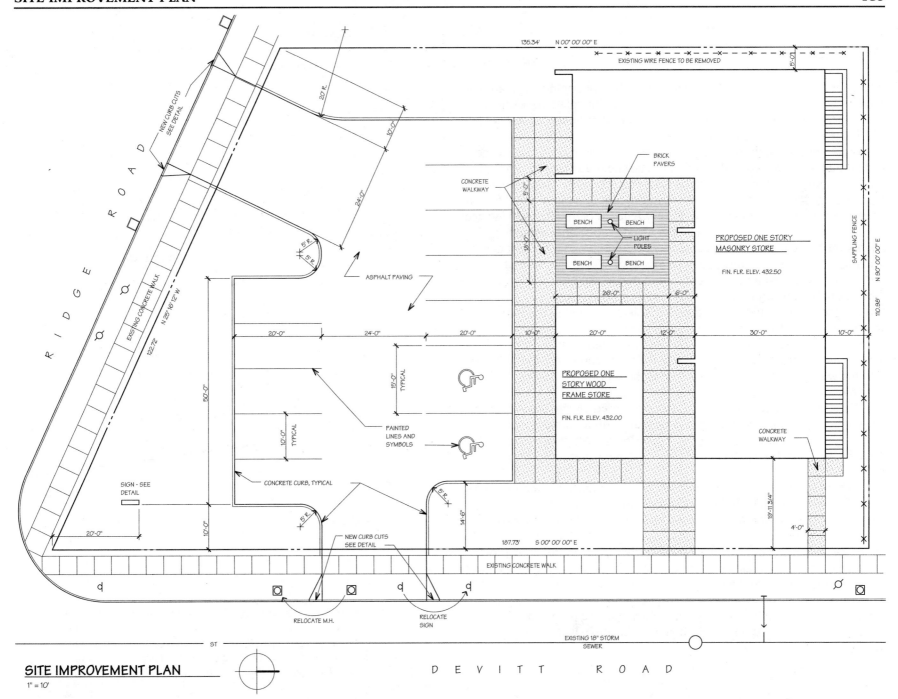

## SITE IMPROVEMENT PLAN

1" = 10'

RIDGE ROAD

DEVITT ROAD

135.34'   N 00° 00' 00" E

EXISTING WIRE FENCE TO BE REMOVED

NEW CURB CUTS
SEE DETAIL

20' R.

10'-0"

24'-0"

5' R.

5' R.

EXISTING CONCRETE WALK

N 25° 16' 12" W

122.72'

50'-0"

10'-0"

20'-0"

SIGN - SEE
DETAIL

20'-0"          24'-0"          20'-0"          10'-0"

15'-0"
TYPICAL

10'-0"
TYPICAL

PAINTED
LINES AND
SYMBOLS

CONCRETE CURB, TYPICAL

5' R.

5' R.

14'-6"

NEW CURB CUTS
SEE DETAIL

RELOCATE M.H.

RELOCATE
SIGN

187.73'   S 00° 00' 00" E

EXISTING CONCRETE WALK

EXISTING 18" STORM
SEWER

ASPHALT PAVING

CONCRETE
WALKWAY

5'-0"

18'-0"

BRICK
PAVERS

BENCH    BENCH

LIGHT
POLES

BENCH    BENCH

26'-0"          6'-0"

PROPOSED ONE STORY
MASONRY STORE

FIN. FLR. ELEV. 432.50

SAPPLING FENCE

N 90° 00' 00" E

110.98'

5'-0"

20'-0"          12'-0"          30'-0"          10'-0"

PROPOSED ONE
STORY WOOD
FRAME STORE

FIN. FLR. ELEV. 432.00

CONCRETE
WALKWAY

19'-11 3/4"

4'-0"

ST

## Drainage

Piping for drainage on a site is often drawn on the plumbing site plan. For some projects, however, it is included on the civil/site plan. An example is shown here for its own sake but also as an introduction to the manner in which site utilities are delineated.

The architectural drawings show all the components of the drainage system that are attached to the building. The site drawings pick up the system where it reaches the ground as one can see at the lower corners of the buildings on the facing page. Water (from rain or melted snow) from a flat roof on a building comes down in *roof leaders* (vertical pipes) located either inside the building or outside. For sloped roofs, *gutters* and *downspouts* are commonly used. Gutters are horizontal trays that collect water as it comes off the low end of a sloped roof. They are pitched slightly to lead the water to one end where the downspouts are located. A downspout is simply a roof leader. The water from the roof leaders usually goes into an underground pipe that is connected to the *storm sewer* under the street, or it is retained on site.

Water from paved surfaces is collected in underground concrete boxes called *catch basins*. They, too, have pipes connected to them and lead the water to the same storm sewer as the roof water. All this work is shown on the site plan.

Water on turf areas does not have to be carried off site but it may need to be moved. Pipes may need to be placed under driveway entrances if the adjacent ground is low. *Yard drains*, which are essentially catch basins, may be placed in low spots to drain water off to another location.

All of these objects are drawn as symbols. The most common symbols are shown on page 180. Underground pipes are drawn as single thick lines with intermittent arrows showing the direction of flow. Their size and material are noted.

Since these drainage systems flow by gravity, the drafter must provide critical elevations such as top of grate or invert. A *grate* is the metal cover level with the surface that allows water to flow through it into an underground receptacle. *Invert* is the elevation of the bottom inside surface of a pipe. By comparing inverts one can see whether the proper slope has been achieved. Many drafters also note the slope to which the pipe needs to be laid.

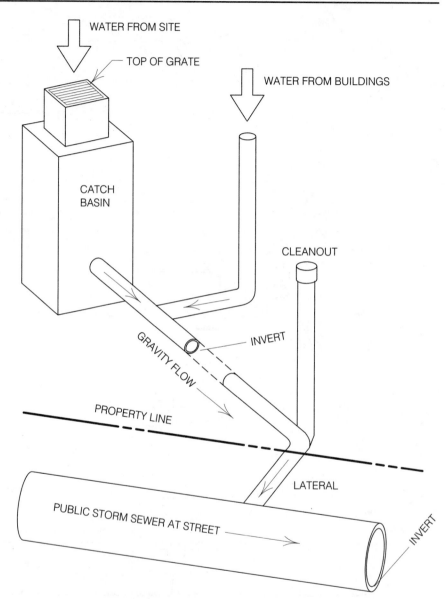

The schematic above shows the basics of a storm water system. Catch basins collect water from large paved areas such as parking lots. Water from the roofs of buildings also goes underground in pipes. Usually the two converge to one larger pipe to be carried to the storm sewer at the street. Wherever a bend occurs, a cleanout is provided for cleaning the lines. The pipe from the property line to the public sewer is called a **lateral**.

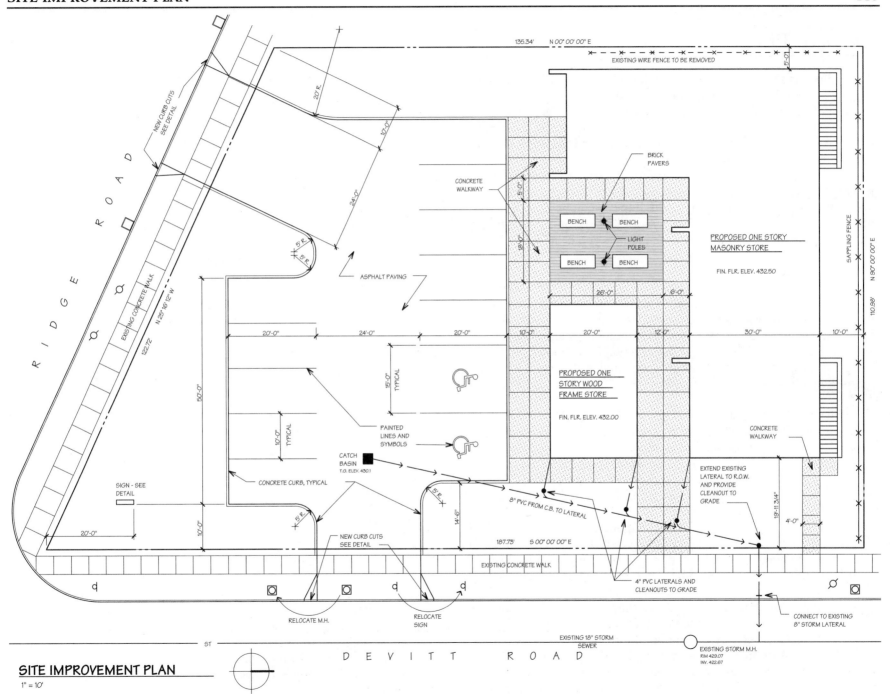

## SITE IMPROVEMENT PLAN

1" = 10'

# GRADING PLAN

The grading plan shows the earth work that must be done to accommodate the project. This is done by drawing existing and proposed contours and spot elevations. The primary purpose of site grading is usually to manage water. The contours are established to lead water to catch basins, to keep it from running onto adjacent properties, and sometimes to hold it on the property temporarily.

Existing contour lines may be transferred from the site survey onto the grading plan. Remember that contour lines are drawn as broken lines of thin weight and that every 5th or 10th line is medium. The elevation numbers are noted adjacent to each contour line just outside the property line. For large properties another set of numbers can be added in between. The numbers themselves can be small in size: 1/8" or 3/32" is usually adequate.

Proposed contour lines are drawn as continuous lines the same weight as the existing contour lines. The proposed contours must diverge from and return to the existing contour lines within the property. If they did not, it would mean that grading work would be occurring on the adjacent parcel of land, and that is not acceptable. As shown on the facing page, the new contour lines are numbered in the center and are enclosed in a box, thereby distinguishing them from existing contour lines.

While contour lines show the land's slope in general terms, spot elevations establish heights at critical points. Existing spot elevations are shown the same way as on the survey map.

Proposed spot elevations are distinguished differently. The most common ways of showing proposed spot elevations are shown on this page. The only other information that is transferred from the site survey is the location of any benchmarks that may be on the property.

As in the site improvement plan it is important to show the building footprint, title it, and—most importantly—give its first floor elevation. The floor elevation should be higher than the surrounding ground to keep water from running into the building. Good design also includes sloping the ground away from the building to lead water away.

One final point about contours is that they should be perpendicular to paved areas. If they are not, then the paved area will be tilted to one side or the other. This is very evident on narrow paved areas such as sidewalks. On larger areas such as parking lots this principle generally applies but must be modified to lead water out of corners and towards catch basins. This can be seen in the grading plan on the facing page.

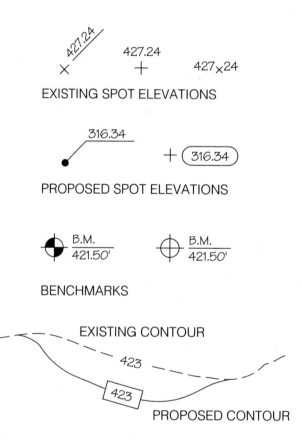

EXISTING SPOT ELEVATIONS

PROPOSED SPOT ELEVATIONS

BENCHMARKS

EXISTING CONTOUR

PROPOSED CONTOUR

*The symbols shown here are fairly common but this is one area where you are likely to see many variations. What is important is to be consistent; adopt what is used in your office or area and stay with it. In this case, a legend on the site plan would be a big help to someone reading the drawing.*

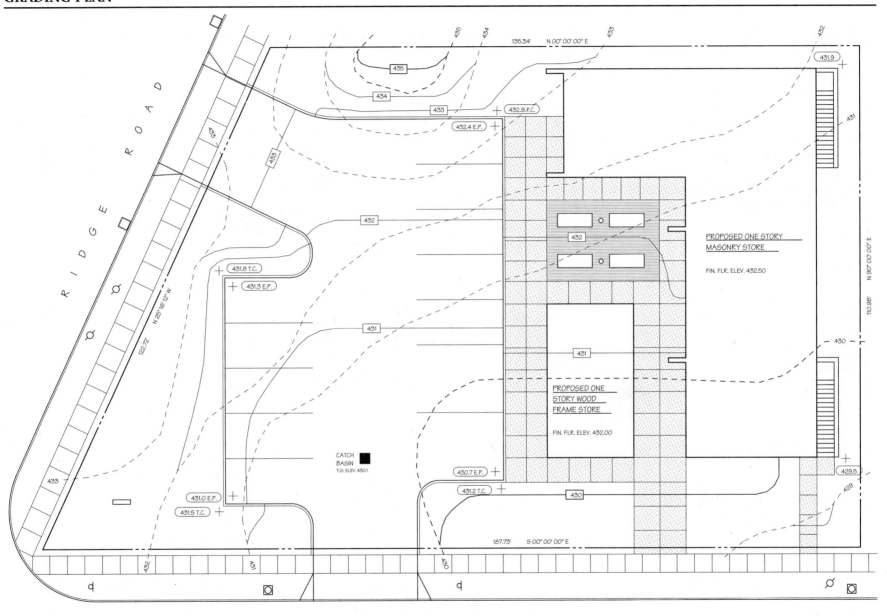

435
434
433
435
434
135.34'   N 00° 00' 00" E
433
432

431.9

432.9 T.C.

432.4 E.P.

432

PROPOSED ONE STORY
MASONRY STORE

FIN. FLR. ELEV. 432.50

431

431.8 T.C.

431.3 E.P.

432

431

431

PROPOSED ONE
STORY WOOD
FRAME STORE

FIN. FLR. ELEV. 432.00

430

N 90° 00' 00" E

110.98'

433

CATCH
BASIN
T.G. ELEV. 430.1

430.7 E.P.

431.2 T.C.

430

429.5

429

431.0 E.P.

431.5 T.C.

187.73'   S 00° 00' 00" E

RIDGE ROAD

N 25° 16' 12" W

122.72'

432

431

430

GRADING PLAN

1" = 10'

DEVITT   ROAD

# LANDSCAPING PLAN

The landscaping plan is conceptually the same as the site improvement plan except that all the elements shown on the plan are natural instead of manmade. The landscaping plan shows shrubs, planting beds, trees, and grass areas. Like the site improvement drawing, reference must be made to which items are existing (and either to be removed or to be left alone) and which are proposed.

Lawn areas usually constitute the largest area of landscaping work. Planting beds are generally the next

largest area to be drawn, and they are done by showing individual plants. Schedules of plant materials accompany the plan and indicate size and species of plant. For small plants an "on center" (o.c.) dimension is given; for larger plants and trees the number of plants is given. Because this information is given in the planting schedule, no dimensions are needed on the drawing.

All plant materials are assumed to be new unless otherwise noted. In the example on the facing page, notice that three existing trees are to remain. They are so identified by their symbol, but are also noted to remain. One existing tree is shown and labeled to be removed. The new plants are noted with a symbol and tied to the planting schedule. A continuous leader line connects the plants to the symbol, leaving no question as to which species each is.

There is no specific hierarchy to the lettering on a landscape plan; thus it can all be done between 3/32" and 1/8" depending on space. Lettering in the schedules may be larger than in the general notes.

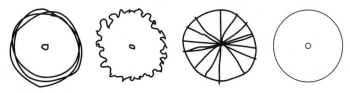

DECIDUOUS TREES

DECIDUOUS OR EVERGREEN          EVERGREEN TREE

SMALL TREES AND SHRUBS

SMALL PLANTS AND GROUND COVERS

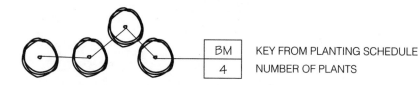

| | BM |   | KEY FROM PLANTING SCHEDULE |
| | 4 |   | NUMBER OF PLANTS |

USE OF SYMBOLS AND TIE-IN TO PLANTING SCHEDULE

| PLANTING SCHEDULE | | | | | |
|---|---|---|---|---|---|
| NO | KEY | BOTANICAL (LATIN) NAME | COMMON NAME | CALIPER | ROOT |
| 12" O.C. | AR | AJUGA REPTANS | BUGLEWEED | | C |
| 11 | BM | BARBERIS MENTORENSIS | MENTOR BARBERRY | 1" | C |
| 5 | CA | COTONEASTER APICULATUS | CRANBERRY COTONEASTER | 1/2" | C |
| 1 | CF | CORNUS FLORIDA | FLOWERING DOGWOOD | 2" | BB |
| 5 | EA | EUONYMUS ALATA | WINGED EUONYMUS | 1" | BB |
| 2 | PT | PINUS THUNBERGIANA | JAPANESE BLACK PINE | 3" | BB |
| 6 | RA | RHODODENDRON ATLANTICUM | COAST DWARF RHODODENDRON | 1" | C |
| 3 | SV | SYRINGA VULGARIS | COMMON LILAC | 1" | C |

C=CONTAINER  BB=BALL & BURLAP  BR=BARE ROOT

*The plants, symbols, and schedule all work together to communicate the needed information. Notice that the **botanical** (Latin) name is given for each plant. While the common name may vary from region to region, the botanical name is the same almost everywhere in the world.*

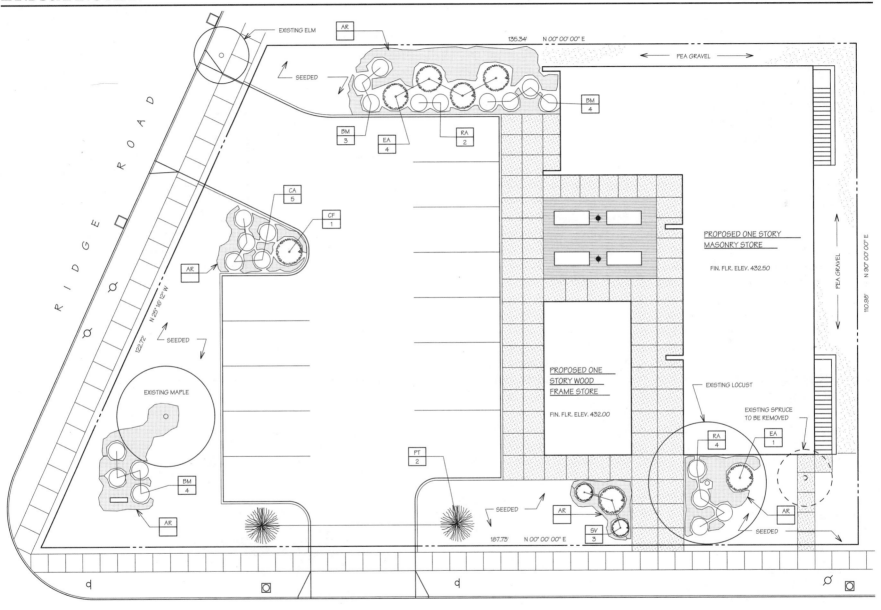

LANDSCAPING PLAN

1" = 10'

# DETAILS

Many site details are stock details that are adapted for a specific project with minor adjustment. There are few, if any, details drawn completely new for a project. Notice that many of the details drawn here are marked **N.T.S.** or <u>N</u>ot <u>T</u>o <u>S</u>cale, showing their multi-purpose nature.

Site details are generally sections and are drawn at architectural scales between 1/2" and 1-1/2". Sometimes areas of the plan need to be enlarged into a detail to a scale of 1/4" to 1/2".

Where possible, details are fit on the same sheet as their related plan, whether it be site improvement, grading, or landscaping. For larger projects there are enough details to create separate sheets for them.

The sequence for details has already been outlined in this text. The basic steps are as follows:

- layout the basic shape and features
- draw the objects with object lines, hidden lines, etc.
- add dimensions and notes
- add reference symbols
- add material indication lines or patterns
- render thick and very thick lines their proper thickness

Rather than illustrating this procedure again, several typical site details are presented here to show their nature and the variety they represent.

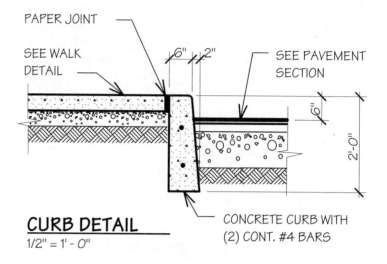

## CURB DETAIL
1/2" = 1' - O"

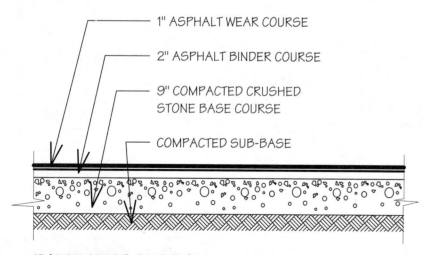

## PAVEMENT SECTION
1/2" = 1' - O"

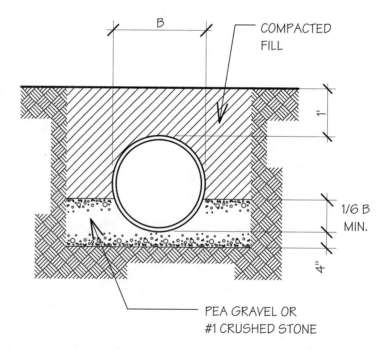

## TYPICAL BEDDING DETAIL
NTS

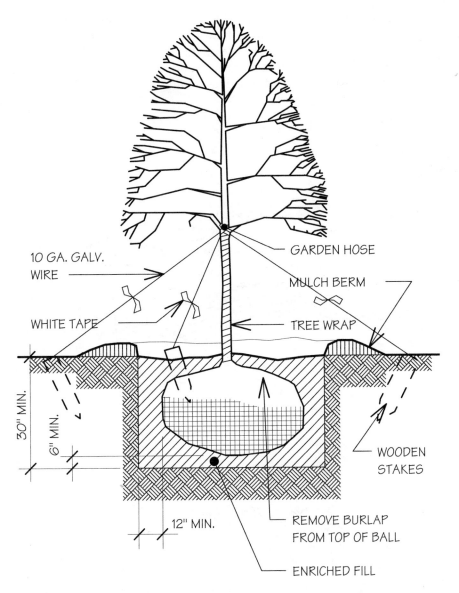

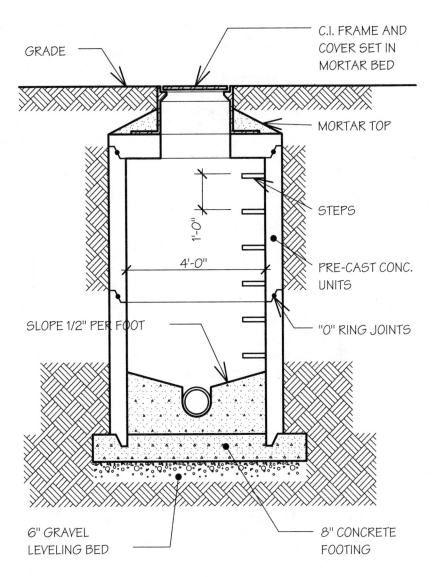

10 GA. GALV. WIRE

GARDEN HOSE

MULCH BERM

WHITE TAPE

TREE WRAP

30" MIN.

6" MIN.

12" MIN.

WOODEN STAKES

REMOVE BURLAP FROM TOP OF BALL

ENRICHED FILL

## TREE PLANTING DETAIL
NTS

GRADE

C.I. FRAME AND COVER SET IN MORTAR BED

MORTAR TOP

1'-0"

4'-0"

STEPS

PRE-CAST CONC. UNITS

SLOPE 1/2" PER FOOT

"O" RING JOINTS

6" GRAVEL LEVELING BED

8" CONCRETE FOOTING

## CATCH BASIN/MANHOLE DETAIL
NTS

# KEY TERMS

| | |
|---|---|
| profile | yard drain |
| footprint | grate |
| roof leader | invert |
| gutter | lateral |
| downspout | botanical name |
| storm sewer | NTS |
| catch basin | |

# PRACTICE EXERCISE

Using the site survey provided on this page, create the following:

- site improvement plan
- grading plan
- landscaping plan
- 2 site details

Place a building on the site that is 40' wide and 60' deep. It is to be a one story masonry building. Zoning requires setbacks of 80' from the front and 20' sides and rear. Assume a flat roof with one roof leader coming out from one corner of the building.

A one story, wood frame building that is 20' square is to be located on the site away from the masonry building but adjacent to the driveway. It has a hip roof with gutters and downspouts.

Add a driveway and parking for eight vehicles. The driveway should be 20' wide, and parking spaces 10' x 20'. Add a sidewalk from the parking lot to the building. Include an outdoor seating area in front of the building. Select materials common in your area.

Grade the property to standards presented in this chapter. Add eight to ten plant species and create a landscaping layout. Consult a gardening catalog to select your plants.

Finally, draw two site details of your choice; you may redraw them from this chapter, find others in reference books, or create your own.

Use media as directed by your instructor. Create your own project title and develop a title block to use on all your drawings. Number and cross reference your sheets according to standards previously introduced.

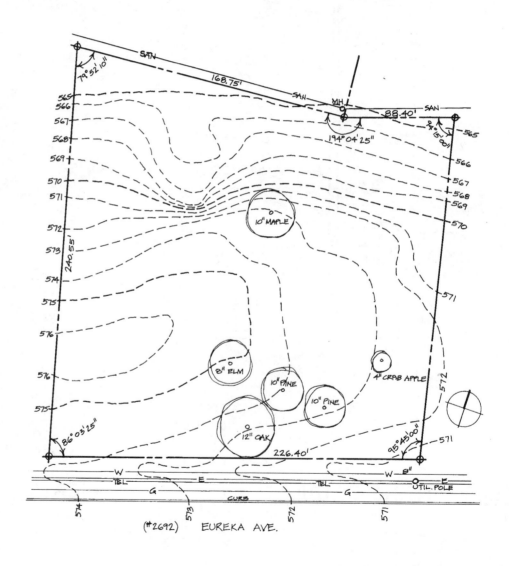

# 15

# CONSTRUCTION DRAWINGS

## • ARCHITECTURAL

### OBJECTIVES

*By the end of this chapter students should be able to, given a building design, draw basic:*

- floor plans
- room finish and door schedules
- building and wall sections
- exterior and interior elevations
- details

FIRST FLOOR PLAN

We have seen that civil/site drawings define the work that must be done to the land surrounding a structure. The structural drawings will define the parts of the building that support all the other materials.

We now focus on architectural drawings. These drawings define the building envelope, its interior and exterior finishes, and most of the remaining built-in features.

Architectural drawings are the most complex because they must define more components than any of the other disciplines. All other disciplines are supportive in a sense because the architectural design is most visible. The other systems are designed to fit within the architectural design.

The architectural staff leads the way in most building construction projects. It is predominantly their responsibility to see that their drawings are coordinated with all disciplines.

Given the importance and complexity of the architectural design, it is understandable that there are more drawings found in this discipline than any other.

# FLOOR PLANS

For a project involving a building, especially a habitable one, there is no more important drawing than the floor plan. This is the drawing that almost always explains the project most clearly. It is usually the drawing begun first. In fact, almost every other drawing springs from the floor plan.

## Wall Layout

In general, floor plans are drawn to a scale large enough to be able to show all the necessary information clearly. The following guidelines are generally accepted standards:

| | |
|---|---|
| small scale buildings such as houses | 1/4" = 1' - 0" |
| large scale buildings | 1/8" = 1' - 0" |
| enlarged plans to show detail | 1/2" = 1' - 0" |

Occasionally a building is designed such that it will not fit onto a standard size sheet even if it is drawn at one of these scales. In this situation the decision is usually made to use one of the less common scales of 3/16" = 1' - 0", 3/32" = 1' - 0", or 1/16" = 1' - 0".

The walls are drawn with layout lines at this point since all the openings are not yet located.

## Openings

With all of the walls laid out, the next step is to draw the openings. You may continue to use a hard lead and place small layout lines to mark the openings.

Openings include all of the following and others which may be found in special situations:
- windows
- doors
- archways
- pass throughs
- louvers and vents
- access openings

Objects built into the wall but not penetrating it, such as medicine cabinets and electrical panels, are treated differently and need not be drawn at this time.

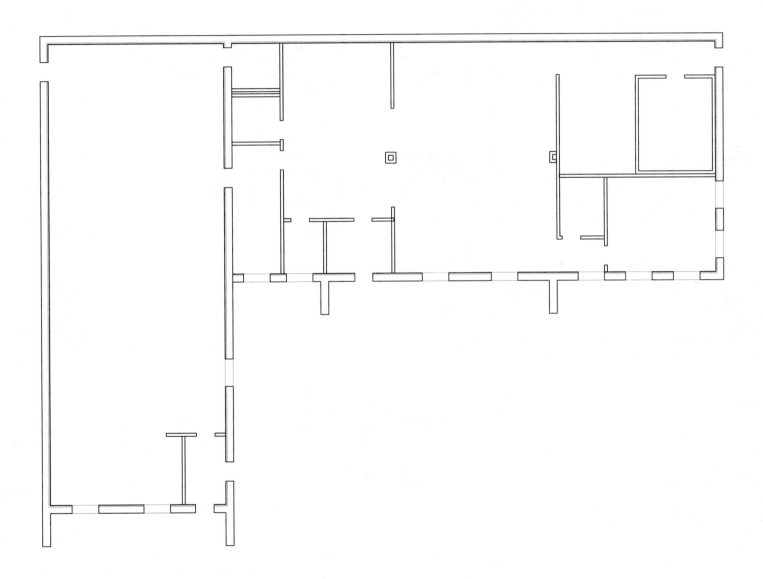

## Windows

Windows at small scale are drawn with thick or medium lines whether single, double, or triple glazed. All operable windows are shown closed. The interior and exterior edges of the sill are drawn as thin lines flush with the wall surfaces or with actual overhang at larger scales.

At larger scales, thick lines are used to draw the exposed faces of the frame. The larger the scale, the more accurately shaped the frame can be drawn.

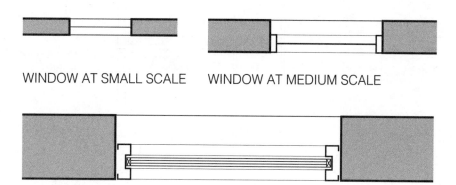

WINDOW AT SMALL SCALE    WINDOW AT MEDIUM SCALE

WINDOW AT LARGE SCALE

## Doors

Door frames are drawn the same as window frames except they appear in their open positions. Swing doors usually open 90° and are drawn as such. If they swing beyond 90° this is shown with the door swing and not the door.

Since doors are in section, they are drawn with thick lines, and since they are only about 1-3/4" thick, a single thick line is drawn. The door swing is really not an object but rather an indicator and thus is drawn as a thin phantom line.

# Design Elements

Anything that is attached to the building and visible on the plan is drawn at this time. This is assuming that it is part of the project and therefore the contractor's responsibility to provide and install.

Occasionally it is desirable to draw an item that is not part of the project, just to help explain the design. A refrigerator is a good example of this. Even though the contractor would not be supplying the refrigerator, it helps him or her to see why there is a gap in the cabinets and to allow for placement of the appliance. It is important to clearly note that the refrigerator will be provided by and/or installed by someone else.

There are countless elements that you may find in any number of special building types. Following are the most common elements that are included on floor plans:

- casework
- counter tops
- specialties such as chalkboards, tackboards, compartments, vents, fireplaces, lockers, mail boxes, partitions, scales, sun control devices, phone booths, and bathroom accessories such as towel bars, paper dispensers, etc.
- equipment such as vaults, vending machines, and special items for darkrooms, churches, schools, restaurants, medical buildings, theaters, etc.
- built-in furnishings such as cabinets, seating, and artwork
- conveying systems such as stairs, elevators, escalators, lifts, dumbwaiters, hoists, cranes, and material handling systems
- plumbing fixtures such as tubs, showers, toilets, sinks, etc.
- mechanical equipment such as HVAC units, controllers, etc.
- electrical equipment: lighting, power units, communication equipment, etc.

The larger the scale, the more detail you will be able to show for each of these items. However, some items need not be shown in too great detail because they are manufactured items. Towel bars and sinks, for example, are basically defined in the specifications. They are stock goods that are shipped to the site and need only to be installed by the contractor. The drawings merely show where to put them.

Custom wood stairs, on the other hand, are built by the carpenter on the site. They will require much more detail. For a feature as complicated as a staircase, enlarged plans and details are needed.

In general, use medium lines to outline all of these components. This is because they are not in section and are beyond the section line. For items above the section line such as upper kitchen cabinets, use hidden lines to outline the object.

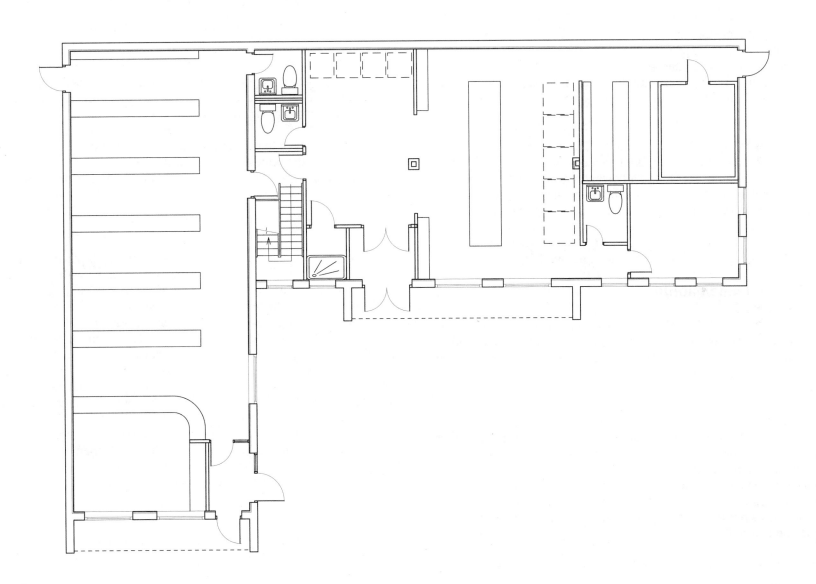

# Dimensions

There are three major types of building systems and as a result there are three methods of dimensioning. You must apply each one to its proper situation. They are:
- light frame (wood and steel stud)
- masonry (brick, block, and tile) or cast-in-place concrete
- heavy frame (structural steel and heavy timber)

Sometimes there is a mixture of these building systems on a given floor plan, so as many as three will be used. This may complicate the drawing and challenge the drafter, but taken one at a time it is possible to sort through the complexity. On the following pages we will address each procedure individually. Just remember that it is quite possible that you will need to use more than one procedure on a single floor plan.

For the general mechanics of size, line thickness, and spacing of dimensions, see the earlier section on dimensions. Presented here are guidelines specific to floor plans on architectural working drawings.

## Light Frame Dimensioning

There are basically four "fields" for dimensions on a floor plan that must be completed. It will require two or more strings of dimensions to adequately cover each of these, so you will eventually have twelve or more sets of dimensions on your plan. The four fields are as follows.

### Overall

These two dimensions are used to establish the overall length and width of the plan. One is placed toward the top of the sheet and the other toward the left. If, however, there are offsets only toward the bottom or the right, it is common to put the overall dimensions on these sides of the plan instead.

Some offices put the overall dimension on both sides to avoid any possible confusion.

### Offset

Any side that has an offset (a protrusion or recess) needs an offset dimension line. One string of dimensions may be used for all offsets on each side.

The example on the facing page has no offsets; therefore it has no offset dimensions. The other building on the site does illustrate this concept and will be shown next.

### Openings

All exterior windows, doors, louvers, etc., are dimensioned with this string. The dimensions are taken to the center line of these openings, which is unique to light frame construction.

The dimensions begin at the **outside** face of one exterior wall, indicate the distance to the **center line** of each opening, and end at the **outside face** of the opposite exterior wall.

If there are no windows in an offset, then no dimension string is placed there as this would only be a repeat of the offset dimension.

### Interior

Each wall, opening, and feature that must be critically placed inside a building needs to be dimensioned. This could be done with one string or ten, since no two buildings are exactly the same. The main guideline is to be efficient and use the fewest lines possible.

Dimension strings generally go entirely across the building, picking up as many walls and openings as possible. Additional strings are added until all openings and walls have been located.

Doors located in the corners of rooms and those centered in walls do not need to be dimensioned because these are standard details. In practice this constitutes the great majority of doors, so actually very few doors need to be dimensioned.

Interior walls and openings are dimensioned to their **center lines**. A common alternate method for walls is to dimension them from face of stud to face of stud. The dimension is the same either way. To avoid confusion use only one method throughout a set of drawings.

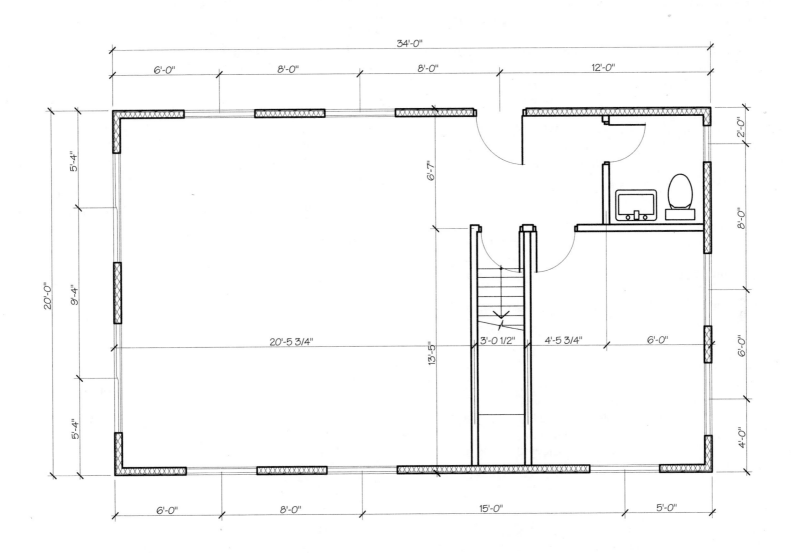

## Masonry Dimensioning

Masonry and cast-in-place concrete buildings are dimensioned exactly the same as light frame buildings except for openings and interiors. The other descriptions are reprinted here for your convenience.

In general, all masonry dimensions should be modular, or a multiple of the masonry unit being used. For example, a concrete block is 1'-3 5/8" long. The mortar joint between blocks is 3/8", giving a total length of 1'-4". Half sized blocks are available to help turn the corner, so an 8" module is the standard for concrete block.

Designers/drafters should strive to make building dimensions a multiple of this module (2'-0", 2'-8", 8'-0", 15'-4", etc.) in order to avoid excess cutting of block. This will save on labor costs.

### Overall

These two dimensions are used to establish the overall length and width of the plan. One is placed toward the top of the sheet and the other toward the left. If, however, there are offsets only toward the bottom or the right, it is common to put the overall dimensions on these sides of the plan instead.

Some offices put the overall dimension on both sides to avoid any possible confusion.

### Offset

Any side which has an offset (a protrusion or recess) needs an offset dimension line. One string of dimensions may be used for all offsets on each side. As one can see in the example on the facing page, the "L" shape of the building creates an offset in two directions.

The wing walls by the entrance are also considered offsets and have been dimensioned as such.

### Openings

All exterior windows, doors, louvers, etc., are dimensioned with this string. Opening dimensions with masonry and concrete are taken to the outsides of these openings.

The dimensions begin at the **outside face** of one exterior wall, indicate the distance to the **edges** of each opening, and end at the **outside face** of the opposite exterior wall.

If there are no windows in an offset, then no dimension string is placed there as this would only be a repeat of the offset dimension.

Another common practice is to add the initials M.O. after each opening dimension. This is an abbreviation for **masonry opening**. It is intended to clarify that the dimension is not the size of the window or door found in the opening but the size of the opening from face of masonry to face of masonry.

### Interior

Each wall, opening, and feature that must be critically placed inside a building needs to be dimensioned. This could be done with one string or ten, since no two buildings are exactly the same. The main guideline is to be efficient and use the fewest lines possible.

Dimension strings generally go entirely across the building, picking up as many walls and openings as possible. Additional strings are added until all openings and walls have been located. Both wall and opening dimensions are taken from **face to face**. In this way wall thicknesses are defined.

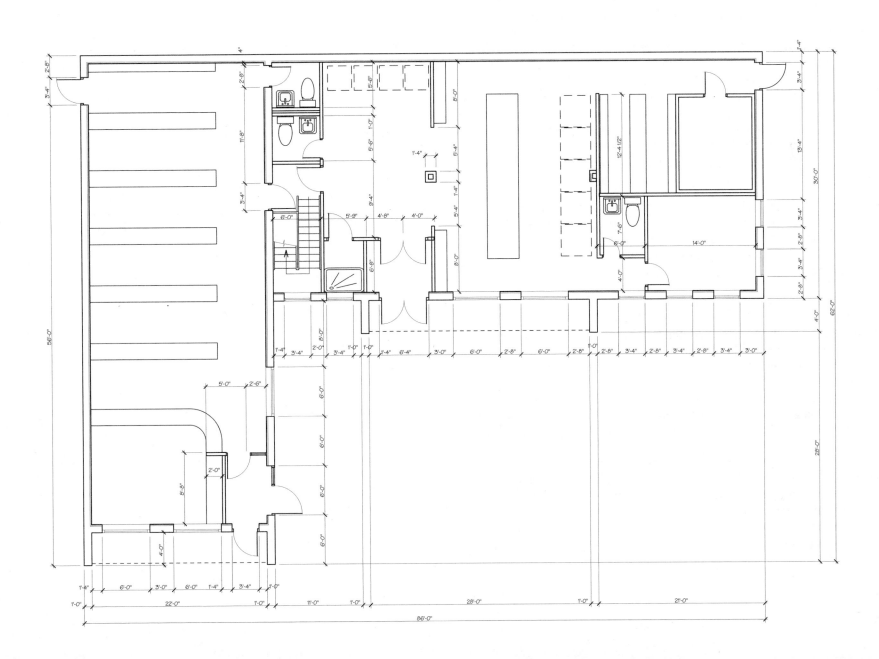

## Heavy Frame Dimensioning

Dimensioning a heavy frame building basically adds one more field of dimensions to the four basic ones already discussed. The five fields are:
- structural grid – provides center to center dimensions of all columns and isolated footings
- overall dimensions – establish the total length and total width
- shape or offset dimensions – define the shape of buildings other than simple squares or rectangles
- opening dimensions – locate all windows, doors, vents, etc.
- interior dimensions – locate all interior walls and openings

### Structural Grid

Although the structural grid is dimensioned on the structural drawings, it is commonly repeated on the architectural drawings. The reason is to relate the walls to the grid. Since the structural framework will already be in place when the walls go up, it becomes the reference from which the walls are built.

### Overall, Shape, Openings, Interior

After the structural grid is erected, the building is enclosed with some combination of construction types. As a result, the rest of the dimensions on a heavy frame building are a combination of the other dimensioning conventions.

On steel frame buildings, for example, it is quite common to see exterior walls and interior partitions made of both concrete masonry and drywall covered light steel framing. The result is a floor plan with all three dimensioning systems superimposed.

The one important dimension that must be included is one that ties the walls to a grid line. Notice how at various points on the plan shown on the facing page, each wall is dimensioned back to a grid line (which is actually the center line of a column). Without this the contractor has no way of knowing where to start walls in relation to the steel or wood columns that have already been erected.

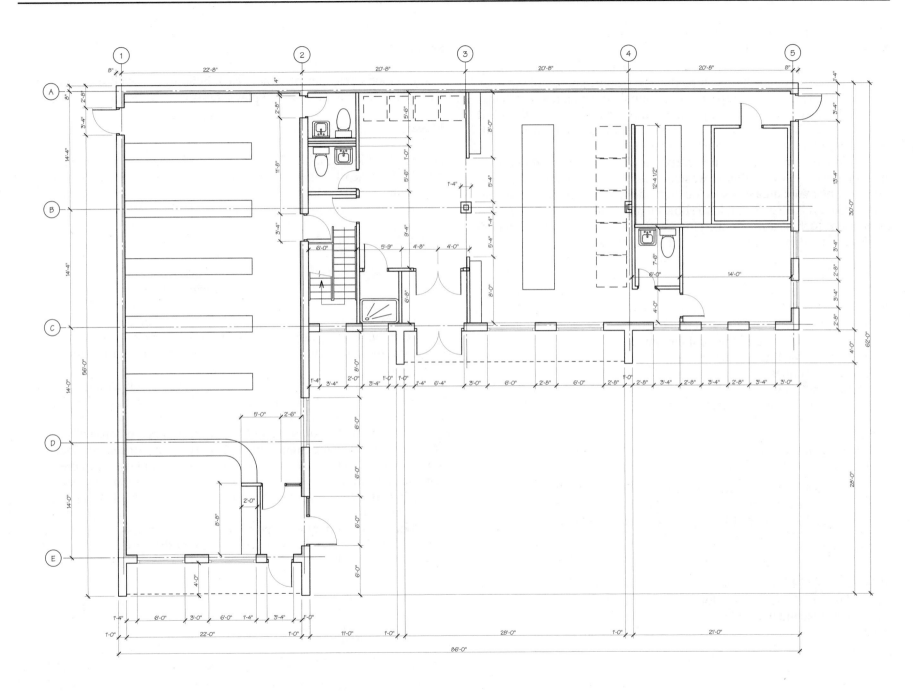

## Material Indication

A small but important step in completing the graphic portion of the floor plan is to indicate, with the appropriate symbol, any material that is in section. Most vertical elements, predominantly walls and columns, need to be pochéd. The appendix has a complete list of material symbols. All material symbols are rendered with very thin lines.

For large areas it is generally acceptable to poché only part of the area; otherwise valuable time is wasted. You can adequately communicate that the entire area is a specific material by showing the poché only periodically. It is also desirable to show all materials that comprise a component such as a wall, which frequently has layers of several materials. Like other drawing parts, this is subject to the scale in use.

## Notes

The next step in drawing a floor plan is to add all the lettering. Much has already been presented in the general chapter on lettering. The following is a checklist guide to lettering for floor plans.
- ❏ drawing title with scale
- ❏ room titles
- ❏ dimensions
- ❏ component identification
- ❏ general notes

## Reference Symbols

Several symbols must be added to the plan to relate it to the other drawings. They are drawn with medium lines unless otherwise noted. They were presented in Chapter 13 but are briefly explained here.

### North Arrow

All plans should have a north arrow. Select one from Chapter 13 that you like and use it. Do not make it too fancy because there are more important things to spend your time on. Align the arrow with and place it near the drawing title.

### Section Marks

Using section lines, indicate where each building and wall section is cut. Do this in the same place on all plans if this is a multi-story building.

### Elevation Symbols

Rooms with a lot of built-in features need interior elevations drawn to explain them clearly. The elevation symbol is placed in the room and arrows are numbered indicating which elevations are drawn. The sheet number where they can be found is located in the center.

### Room, Door, and Window Numbers and Letters

Each of these has its own shape. Room number symbols are placed under the room titles; door numbers are placed within the door and its swing. For a non-swing door, the symbol is placed inside or outside the door.

Windows are usually given letters, and the symbols are placed outside of the building adjacent to the window. However, many offices only show these particular symbols on the exterior elevations, not the floor plans.

Each window type and size gets its own letter. If a building has two windows exactly the same, its letter is actually seen twice on the plan. Windows are lettered A to Z.

Doors and rooms, on the other hand, are numbered consecutively. For first floor rooms the numbering is 101, 102, etc., for second floor rooms, 201, 202, etc., and so on. Small rooms within large ones are commonly given numbers such as 201A, 201B, etc. Numbering begins at the main entrance and proceeds around the plan and back to the beginning again.

Doors are most frequently given the same number as the room they are found in. For example, door 207 is found at the entrance of room 207, and door 207A is the closet door in that room.

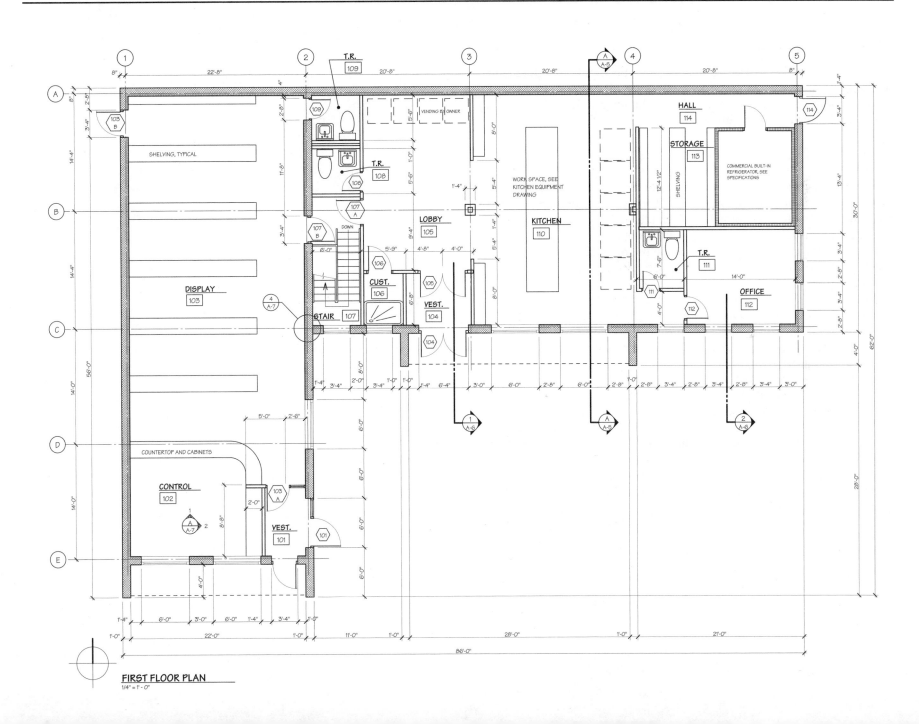

**FIRST FLOOR PLAN**
1/4" = 1' - 0"

# SCHEDULES

There are several types of schedules that may be found on architectural working drawings. Two in particular are almost always found and are quite important to understand. The first is the room finish schedule, and the second is the door schedule.

## Room Finish Schedules

As we have just seen, all rooms on the floor plan are given names and numbers. This is done to cross reference them to the schedules. In fact, these are the first two columns in the schedule.

There are usually three major groups of columns that follow: floor, wall, and ceiling. The floor columns name the material and sometimes the finish to be used. A related column listing the base is also usually part of this group.

The group of columns that list the walls can be more complicated only because there are usually four walls and frequently their finishes vary within a room.

Four (or three or five if an odd shaped room) sub-columns may be established to list the finishes of each wall. They are usually named north, south, east, and west. Notice here that it is common for more than one material to be listed or checked. This is because walls such as those in bathrooms often have more than one finish material on them, i.e., tile from floor to 4' and drywall above. One type of finish may also need to be further finished. Drywall covered with paint or wallpaper is the most common example.

Ceilings are done much the same as walls except that there is only one of them rather than four. The one unique feature that must be part of a ceiling's description is its height.

Finally, a room finish schedule usually contains a remarks column to identify any odd situation that may apply to a specific room such as "artist mural on north wall" or "moisture resistant drywall walls and ceiling." Basically it is used to note any exceptions or deviations from the information in the other columns.

It is generally a good idea to leave a few extra spaces at the end of each floor listing just in case a closet gets added and you need to add one more room number late in the project.

Numbers are used in the columns on the facing page. Dots and material abbreviations are also very common and may be used.

# ROOM FINISH SCHEDULE

| | | FLOOR | | | | | | BASE | | | | | WALLS | | | | | | CEILING | | | | |
|---|---|---|---|---|---|---|---|---|---|---|---|---|---|---|---|---|---|---|---|---|---|---|---|
| | | QUARRY TILE | CARPET | VINYL | CERAMIC TILE | | | QUARRY TILE | VINYL | CERAMIC TILE | | | CONCRETE BLOCK | DRYWALL | PAINT | CERAMIC TILE | | | EXPOSED | SUSPENDED ACOUSTIC | | HEIGHT | |
| NO. | ROOM NAME | 1 | 2 | 3 | 4 | 5 | 6 | 7 | 8 | 10 | 11 | 12 | 13 | 14 | 15 | 16 | 17 | 18 | 19 | 20 | 21 | 10'-0" | REMARKS |
| 101 | VESTIBULE | 1 | | | | | | 7 | | | | | | 14 | 15 | | | | | 20 | | 9'-0" | |
| 102 | CONTROL | | 2 | | | | | | 8 | | | | | 14 | 15 | | | | | 20 | | 10'-0" | |
| 103 | DISPLAY | | 2 | | | | | | 8 | | | | | 14 | 15 | | | | | 20 | | 10'-0" | |
| 104 | VESTIBULE | 1 | | | | | | 7 | | | | | | 14 | 15 | | | | | 20 | | 9-0" | |
| 105 | LOBBY | 1 | | | | | | 7 | | | | | | 14 | 15 | | | | | 20 | | 10'-0" | |
| 106 | CUSTODIAN | | | | 4 | | | | | 10 | | | | 14 | 15 | | | | 19 | | | | |
| 107 | STAIR | | | 3 | | | | | | | | | 13 | 14 | 15 | | | | 19 | | | | |
| 108 | TOILET ROOM | | | | 4 | | | | | 10 | | | 13 | 14 | 15 | 16 | | | | 20 | | 8-6" | TILE TO +4' |
| 109 | TOILET ROOM | | | | 4 | | | | | 10 | | | 13 | 14 | 15 | 16 | | | | 20 | | 8'-6" | TILE TO +4' |
| 110 | KITCHEN | | | 3 | | | | | 8 | | | | | 14 | 15 | | | | | 20 | | 10'-0" | |
| 111 | TOILET ROOM | | | | 4 | | | | | 10 | | | | 14 | 15 | 16 | | | | 20 | | 8'-6" | TILE TO +4' |
| 112 | OFFICE | | 2 | | | | | | 8 | | | | | 14 | 15 | | | | | 20 | | 9'-0" | |
| 113 | STORAGE | | | 3 | | | | | | | | | | 14 | 15 | | | | | 20 | | 10'-0" | |
| 114 | HALLWAY | | | 3 | | | | | | | | | | 14 | 15 | | | | | 20 | | 10'-0" | |
| | | | | | | | | | | | | | | | | | | | | | | | |
| | | | | | | | | | | | | | | | | | | | | | | | |
| | | | | | | | | | | | | | | | | | | | | | | | |
| | | | | | | | | | | | | | | | | | | | | | | | |

## Door Schedules

Most door schedules start out with the basic information of door number, width, height, and thickness. If there are two doors, the word "pair" is usually printed after the door number or after the basic dimensions. The size given is for one door.

Other basic door information is usually grouped into a series of columns that include the door type (plain, with glass, louver, etc.), material (wood, aluminum, solid or hollow core), and finish (stained, painted, anodized, etc.).

The door type is very often a number that corresponds with a small scale drawing of exactly what the door looks like. This series of door drawings is part of the schedule and is usually found at the bottom or next to the schedule proper, but certainly on the same sheet. These drawings show sizes and locations of glass, louvers, etc. The one shown at right accompanies the schedule on the facing page.

Next the drafter must describe the frame that surrounds the door. A group of columns is created for this purpose. Generally the frame (or buck) type, material, and finish are provided. As with door type, the frame type is usually a letter that corresponds to a detailed drawing. The material and finish are either entered or checked in the appropriate column.

Other columns define various door and frame elements. The ones in our example are common. **Thresholds** are placed below the door between the bottom ends of the frame. On exterior doors they help seal the air gap found at that location. On interior doors they are sloped as a transition between rooms that have different floor levels.

**Weather-stripping** is a thin vinyl or other material applied around the frame. When the door is closed against it, it compresses and seals against excessive energy loss. It is for exterior doors.

An **undercut** is for interior doors. It is an exaggerated gap between the door and floor to allow air to pass more freely between rooms. It is used for rooms whose doors tend to remain closed and have no other ventilation. A **louver** is a grille cut into the door for much more positive air flow.

A last bit of information for doors is its **label** or fire rating. A column to enter either the fire rating in hours or letter is therefore almost always found in the schedule.

As in the room finish schedule, a final column for remarks is reserved for exceptions. Again it is advisable to leave a few blank spaces after each group of doors representing one floor.

On more elaborate schedules you may find columns that list offset for frames, manufacturer's name, hardware type, glass type, glass area, etc. For the beginning drafter it is nice to keep this in mind but a grasp of the basics is more important.

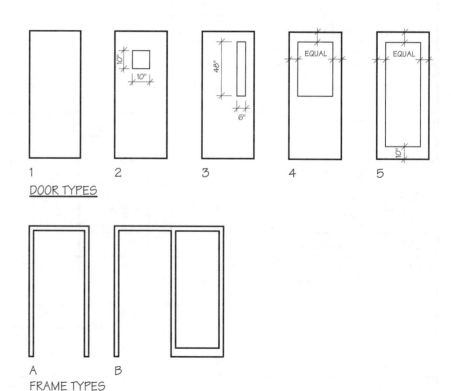

## DOOR SCHEDULE

| NO. | DOOR SIZE | MATERIAL | TYPE | FRAME MATERIAL | TYPE | THRESHOLD | WEATHERSTRIPPING | UNDERCUT | HARDWARE | LABEL | REMARKS |
|-----|-----------|----------|------|----------------|------|-----------|------------------|----------|----------|-------|---------|
| 101A | 3'-0 x 7'-0 x 1 3/4" | AL | 5 | AL | A | AL | • | | 1 | | |
| 101B | 3'-0 x 7'-0 x 1 3/4" | AL | 5 | AL | B | AL | • | | 1 | | |
| 103A | 3'-0 x 7'-0 x 1 3/4" | AL | 5 | AL | B | | | | 2 | | |
| 103B | 3'-0 x 6'-8 x 1 3/4" | HM | 1 | HM | A | AL | • | | 3 | | |
| 104 | PAIR 3'-0 x 7'-0 x 1 3/4" | AL | 5 | AL | A | AL | • | | 1 | | |
| 105 | PAIR 3'-0 x 7'-0 x 1 3/4" | AL | 5 | AL | A | | | | 1 | | |
| 106 | 3'-0 x 6'-8 x 1 3/4" | HM | 1 | HM | A | | | | 4 | | 18" x 24" LOUVER |
| 107A | 3'-0 x 7'-0 x 1 3/4" | HM | 2 | HM | A | | | | 5 | B | |
| 107B | 3'-0 x 7'-0 x 1 3/4" | HM | 2 | HM | A | | | | 5 | B | |
| 108 | 2'-4 x 6'-8 x 1 3/4" | WD | 1 | HM | A | MAR | | 1/2" | 6 | | |
| 109 | 2'-4 x 6'-8 x 1 3/4" | WD | 1 | HM | A | MAR | | 1/2" | 6 | | |
| 111 | 2'-4 x 6'-8 x 1 3/4" | WD | 1 | HM | A | MAR | | 1/2" | 6 | | |
| 112 | 3'-0 x 6'-8 x 1 3/4" | WD | 4 | HM | A | | | | 4 | | |
| 114 | 3'-0 x 6'-8 x 1 3/4" | HM | 3 | HM | A | AL | • | | 3 | | |
| | | | | | | | | | | | |
| | | | | | | | | | | | |
| | | | | | | | | | | | |

# SECTIONS

The two types of sections most commonly used on architectural drawings are building sections and wall sections. Building sections are slices through the entire building. Wall sections are larger scale drawings of the wall portion only. Sometimes, on small buildings the drafter can draw an entire building section at a large scale and have it fit onto one sheet. This eliminates the need for wall sections.

Most of the time, however, the drafter must draw a building section to show the general picture and enlarged wall sections for clarification. It is most common to draw building sections at the same scale as the floor plans and to draw wall sections at 3/4" = 1' - 0". Certainly you will find exceptions.

The primary purpose of a building section is to show how the walls are constructed. An integral part of that are the vertical dimensions. Of particular importance is how all the horizontal elements (roof, ceilings, and floors) frame into the wall.

Building sections show reference symbols because they are stepping stones to more detailed drawings such as wall sections.

## Establishing Heights

On a section, the drafter begins by using layout lines to locate all important heights such as finish floor heights, roof height, top and bottom of footings, and in the case of wood construction, top of wall plate.

## Wall Thicknesses

The drafter can now draw any wall that the section mark passes through. Again, in the case of manual drafting, this is done with layout lines. One common aid in setting up a building section is to tape a print of the floor plan directly above and in line with the section. All walls, doors, windows, etc., may then be projected down to create the section.

At this point the drawing will consist of a light layout of the entire section defining all the spaces. In the case of a wall section the major components will be outlined. In essence you have positioned the drawing on the sheet and created "guidelines" from which the drawing can be accurately completed.

## Define Elements

With the height lines and wall thicknesses as a guide the drafter now can outline all major vertical and horizontal elements such as walls, floors, beams, ceilings, roof, footings, grade lines, etc.

All lines may be drawn at their finished thickness at this point. Keep in mind that any object that has been cut through by the section mark should be outlined with a thick or profile line. Objects within that element in section are drawn with medium lines. Elements beyond the section mark such as doors and windows are drawn with thin lines. The finished grade line is usually drawn as an extra thick line.

These guidelines, as usual, are subject to scale and proximity. At 1/8" scale, for example, it is impossible to draw a 2 x 8 wood joist in section with a thick outline. It is equally impossible to draw several thin layers of a material with a series of medium lines. The drafter must therefore temper the general guidelines by keeping the overall objective of clarity in mind.

Drawing the foundation is usually somewhat simple. Footings, foundation walls, and floor slabs are thick enough to be shown accurately at any scale. The adjacent grade line is almost always a step or two below the finished floor, and a small amount of soil around the footing will later be indicated. In other words, it is important to show how the building "plugs into" the ground.

Walls for a small scale building section may be shown as a simple outline whereas a wall section can show every single material.

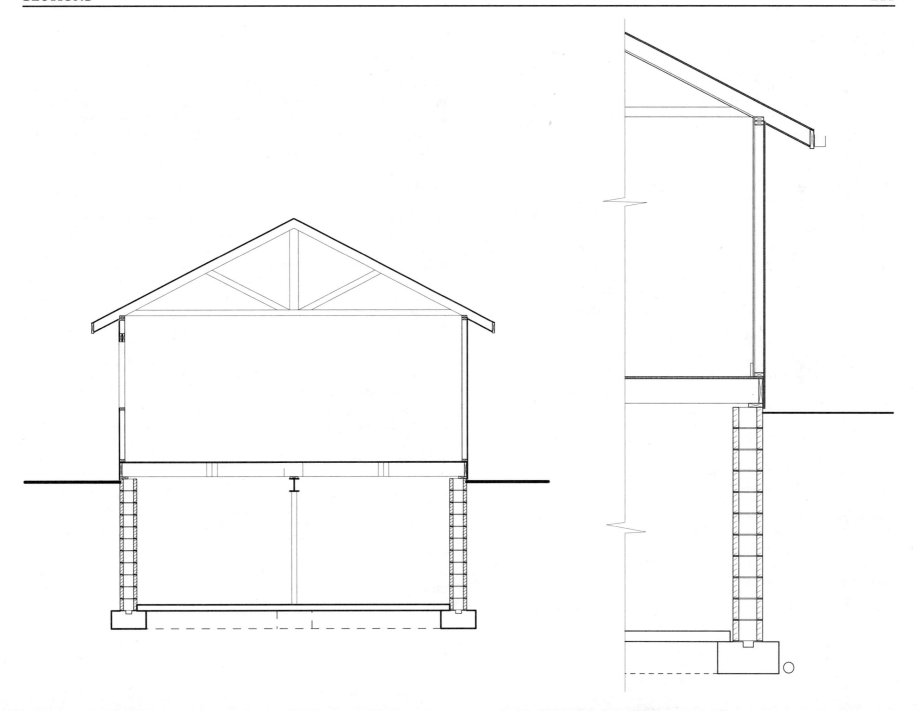

## Dimensioning

The important dimensions to show on a section are the vertical ones, especially floor to floor heights. At least two strings of dimensions are usually required. First the drafter dimensions the floor elevation lines, which were the first lines drawn. These are usually shown on the outside part of the section.

In the case of light frame construction, it is usually more common to dimension to the "top of plate" as that is the critical dimension and the next "platform" begins from that point.

In the case of a concrete slab floor, the drafter usually dimensions to the "top of slab," which is not necessarily the finished floor since tile or carpet will most likely be added.

Whatever the situation, what is important is to letter at the extension line exactly where the dimension is taken to, i.e., "top of sub-floor," "finished floor," "top of beam," etc. The materials listed below normally are dimensioned or have their elevation given as follows:

- masonry – number of courses and dimension
- concrete slab – top of slab
- wood stud wall – top of plate
- wood framed floor – top of (plywood) deck
- steel stud wall – top of studs or channel
- structural steel – top of steel
- steel deck – top of deck

The list above is all that is necessary for exterior dimensions. Dimensions for interior features such as railing heights, stair landings, shelves, etc., are drawn next. As with interior dimensions for plans, careful selection of a string of dimensions can reduce the number you have to create.

Sloped surfaces such as roofs are "dimensioned" using a triangle comparing the rise (vertical) against the run (horizontal). It is a proportion, not a dimension, based on 12. The slope is shown as "x" units of rise for every 12 units of run. Thus a slope of 12:12 is 45 degrees.

Any overhangs or other horizontal dimensions that best show up in section are also dimensioned on sections.

## Material Indication

Indicating materials on a section is done pretty much the same way it is done on floor plans. All are shown as very thin lines; most are in section; and some that are beyond the cut line are drawn if it adds clarity, such as a wood beam beyond. As always, the indication may change depending on the scale used.

## Notes

Lettering on the section primarily calls out the materials shown and explains special installation procedures. Since more information is shown on the wall section, more lettering is possible. Rules for lettering can follow those presented in the Chapter 4.

It is very common to group lines of lettering for materials in a building element if in fact those materials are grouped together. In the example on the facing page, all of the materials that comprise the wall are grouped together, and then only one arrow is needed to show location.

Many times it is not possible to fit all the notes along a single vertical line of lettering. In that situation a second line can be struck. Usually all required notes can fit in this scheme.

Aside from calling out the materials, it may be necessary to provide some notes for installation requirements. Generally, notes are needed in special situations such as non-standard application of materials.

## Reference Symbols

Wall sections have no reference symbols except for their number and the accompanying sheet number they came from located in their title.

On the other hand, building sections are a virtual springboard for larger scale sections and details and can sometimes be covered with reference bubbles. The reference bubble assigns a number indicating the detail or wall section and a sheet number where to find it. The bubble is attached to a circle or oval that defines the area to be detailed elsewhere.

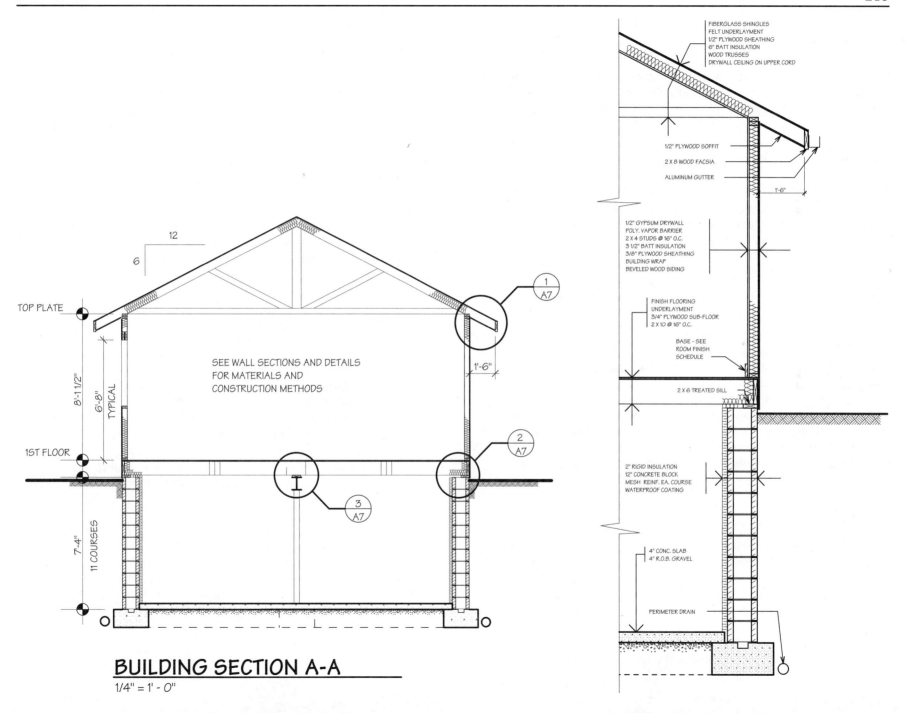

FIBERGLASS SHINGLES
FELT UNDERLAYMENT
1/2" PLYWOOD SHEATHING
6" BATT INSULATION
WOOD TRUSSES
DRYWALL CEILING ON UPPER CORD

1/2" PLYWOOD SOFFIT

2 X 8 WOOD FACSIA

ALUMINUM GUTTER

1'-6"

1/2" GYPSUM DRYWALL
POLY. VAPOR BARRIER
2 X 4 STUDS @ 16" O.C.
3 1/2" BATT INSULATION
3/8" PLYWOOD SHEATHING
BUILDING WRAP
BEVELED WOOD SIDING

FINISH FLOORING
UNDERLAYMENT
3/4" PLYWOOD SUB-FLOOR
2 X 10 @ 16" O.C.

BASE - SEE
ROOM FINISH
SCHEDULE

2 X 6 TREATED SILL

2" RIGID INSULATION
12" CONCRETE BLOCK
MESH REINF. EA. COURSE
WATERPROOF COATING

4" CONC. SLAB
4" R.O.B. GRAVEL

PERIMETER DRAIN

12

6

1
A7

TOP PLATE

SEE WALL SECTIONS AND DETAILS
FOR MATERIALS AND
CONSTRUCTION METHODS

1'-6"

8'-11 1/2"

6'-8" TYPICAL

1ST FLOOR

2
A7

3
A7

7'-4"

11 COURSES

## BUILDING SECTION A-A

1/4" = 1' - 0"

# ELEVATIONS

In the architectural field, elevations as presented in Chapter 8 of this text are an important part of construction drawings. In addition, interior wall surfaces that contain many built-in features are also best explained with what are called interior elevations.

## Exterior Elevations

On the exterior, one elevation is drawn for each face of the building. Thus, a square or rectangular building requires four exterior elevations. Likewise, a pentagon shaped building requires five exterior elevations, and a "U" shaped building needs four plus the two on the inside of the "U" that face inward for a total of six.

Exterior elevations are generally named for the compass direction they face, i.e., "north elevation," "south elevation," etc. If a face is between north and east, one of them is called north to avoid using confusing names such as northeast and southwest elevations.

The primary purposes of exterior elevations are to give a clear picture of what the building looks like and to identify all of the exterior materials. Compared to section drawings, they are less cluttered and usually take less time to complete.

Orthographic projection allows the drafter to use already drawn plans and sections to establish elevations. In fact, taping the plan above and the section to the side is a convenient way of drawing an elevation. The drafter only needs to extend layout lines from these drawings to locate critical points on the elevation. The scale for exterior elevations is, therefore, usually the same as the plan and building section.

## Interior Elevations

Interior elevations are similar to exterior elevations, but differ in the way they are named and/or referenced. The drafter draws interior elevations wherever a wall has features built into it such as kitchens and baths. These rooms have plumbing fixtures, accessories, and cabinets.

Fireplaces, shelving, and equipment rooms are also examples where interior elevations are necessary to show the contractor what needs to be placed and where it needs to be placed.

Interior elevations are sometimes the same scale as the plan if the

plan is drawn at 1/4", but more often they need to be larger. Scales of 3/8" and 1/2" are used most frequently.

### Outlining the Elevation

For exterior elevations, virtually every height is brought over from the section, and every feature is brought down from the plan. Although underground, the foundation is shown as a hidden line. Floor lines are extended even though they are also hidden by wall construction. Roof information, though a hidden line on the plan, is needed to complete the elevation outline.

The only tricky part of exterior elevations is in placing the grade line. That is the one line where the drafter needs to get information from another drawing, namely the site plan. The section and elevation tell the height or elevation of the ground floor, and the site plan tells the height of the surrounding grade with contour lines.

The grade line is established by transferring contour information to the elevation. At the very least, the grade height at each corner should be established and the grade line drawn accordingly. As in the section, the grade line is drawn extra thick. It is often drawn freehand to express its true nature. It is usually the only line on an exterior elevation that is in section.

For interior elevations, the outline is formed by taking a section through the ceiling, adjacent walls, and floor. There is no need to show the adjacent walls in section as that information is found elsewhere. Since this outline represents a section, it is drawn with thick lines.

The one tricky situation for interior elevations is how to draw the side walls when there are cabinets or other features that continue around from the face you are illustrating. Most drafters simply show the feature as an outline.

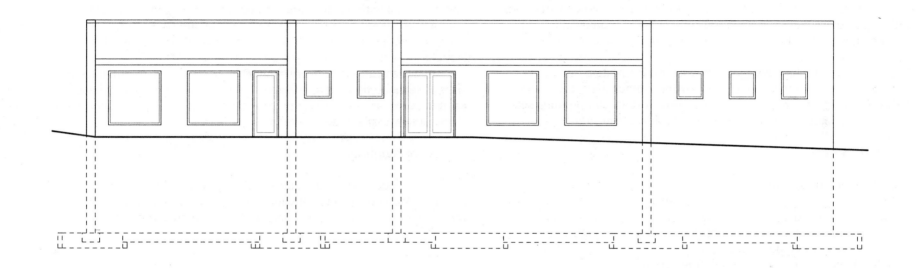

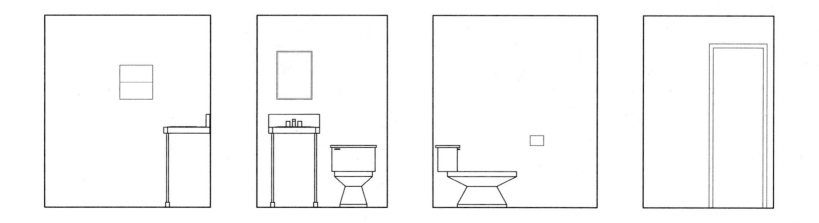

## Dimensioning

Exterior elevations do not have many dimensions. Like building sections, they show floor to floor heights (and top of plate for wood construction) and a slope diagram for sloping roofs. The only other dimension that might be needed is for a projection like a chimney that does not show up elsewhere.

Interior elevations have more dimensions. Every item must be located. Because the plans give the horizontal dimensions, the interior elevations must give the vertical dimensions. Just as with building sections, each element should have its own vertical dimension.

## Material Indication

Indicating materials is an easy task. It is important to realize, however, that material symbols are different in elevation than they are in section. A look at the appendix will show this.

The drafter should try to indicate every material though some are narrow and cannot be pochéd. The only question is when to stop when there is a large area of one material. It is not only boring for a drafter to poché the whole area, it is also a waste of time. Computers can do it with ease and speed, but manually it is usually acceptable to fill only a portion of the area at each end and then leave the rest blank.

## Notes

The only general lettering on exterior elevations is to name each material and point to it with a leader line and terminator. Aligning the names is appropriate, though not doing so is fine. There are so few notes on an exterior elevation that there is little chance that the drawing will look cluttered.

Interior elevations are very similar. In addition it is common to give sizes of stock items such as the note "3024" for a stock kitchen base cabinet 30" x 24". A drawing title and scale is lettered below each drawing as previously explained.

## Reference Symbols

Exterior elevations show all related building section, wall section, and detail marks. They also usually include window symbols and letters. Interior elevations should be referenced to the plan with their title.

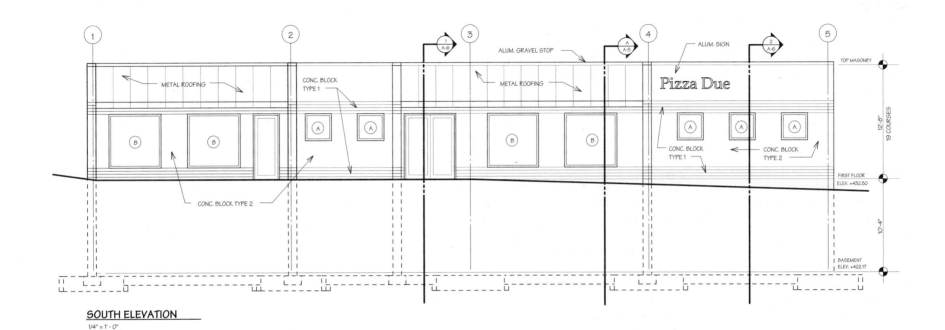

**SOUTH ELEVATION**
1/4" = 1' - 0"

METAL ROOFING · CONC. BLOCK TYPE 1 · ALUM. GRAVEL STOP · METAL ROOFING · ALUM. SIGN · Pizza Due · TOP MASONRY · 12'-8" · 19 COURSES · CONC. BLOCK TYPE 1 · CONC. BLOCK TYPE 2 · CONC. BLOCK TYPE 2 · FIRST FLOOR ELEV. +432.50 · 10'-4" · BASEMENT ELEV. +422.17

PAINTED DRYWALL

TOWEL ROLL

2'-6"

CERAMIC TILE

4'-6"

**ELEVATION A1**

PAINTED DRYWALL

MIRROR

10"

CERAMIC TILE

4'-0" TYPICAL

4'-6"

**ELEVATION A2**

PAINTED DRYWALL

2'-4"

CERAMIC TILE

PAPER HOLDER

3'-0"

**ELEVATION A3**

PAINTED DRYWALL

CERAMIC TILE

**ELEVATION A4**

**INTERIOR ELEVATIONS**
1/2" = 1' - 0"

# DETAILS

There is always the need to draw some parts of the building even larger than the wall sections. These drawings can illustrate certain conditions more clearly. Most of them are enlarged vertical sections but there are also some in plan view.

There is no set pattern for which details will be needed. Some are quite common. Basically the drafter needs to be prepared to draw a detail for whatever condition is not explained clearly enough in other drawings.

The scale of details is variable. Use a scale large enough to clearly show the detail. This can be accomplished at 1/2" = 1' - 0" to full scale.

Although many scales are used, 1-1/2" and 3" scales are most common. The logic seems to be to draw a detail at double the size of a 3/4" wall section. If that is not clear enough, then double it again to 3". Occasionally 3/4" can do the job.

Sometimes if the space on the sheet does not allow these scales, the drop-back scales are 1" and 1/2". However, these scales tend not to be used because the people who read blueprints get used to seeing details at 1-1/2" and 3" and can visualize the details better.

The drafting procedure for details is basically the same as for the "parent" drawing. For architectural drawings, details can be seen as the end of a logical progression. For example, building sections give overall information. Wall sections zero in to show the juncture of major building planes. And details show the connection of individual elements.

Following are details from our example project. The sequence of detailing has already been presented. It is abbreviated here for review.

- layout the basic shape and features
- draw the objects with object lines, hidden lines, etc.
- add dimensions and notes
- add reference symbols
- add material indication lines or patterns
- render thick and very thick lines their proper thickness

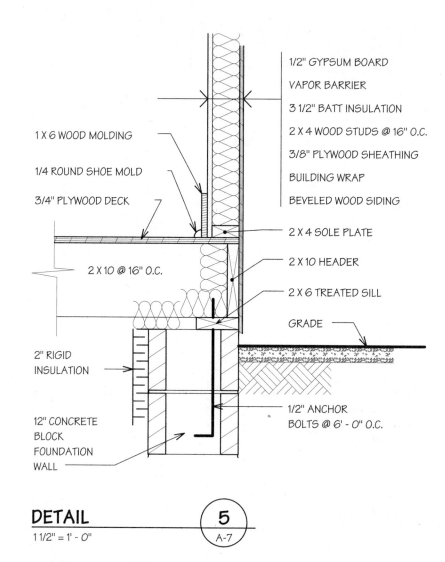

1 X 6 WOOD MOLDING

1/4 ROUND SHOE MOLD

3/4" PLYWOOD DECK

2 X 10 @ 16" O.C.

2" RIGID INSULATION

12" CONCRETE BLOCK FOUNDATION WALL

1/2" GYPSUM BOARD

VAPOR BARRIER

3 1/2" BATT INSULATION

2 X 4 WOOD STUDS @ 16" O.C.

3/8" PLYWOOD SHEATHING

BUILDING WRAP

BEVELED WOOD SIDING

2 X 4 SOLE PLATE

2 X 10 HEADER

2 X 6 TREATED SILL

GRADE

1/2" ANCHOR BOLTS @ 6' - 0" O.C.

**DETAIL**

1 1/2" = 1' - 0"

5

A-7

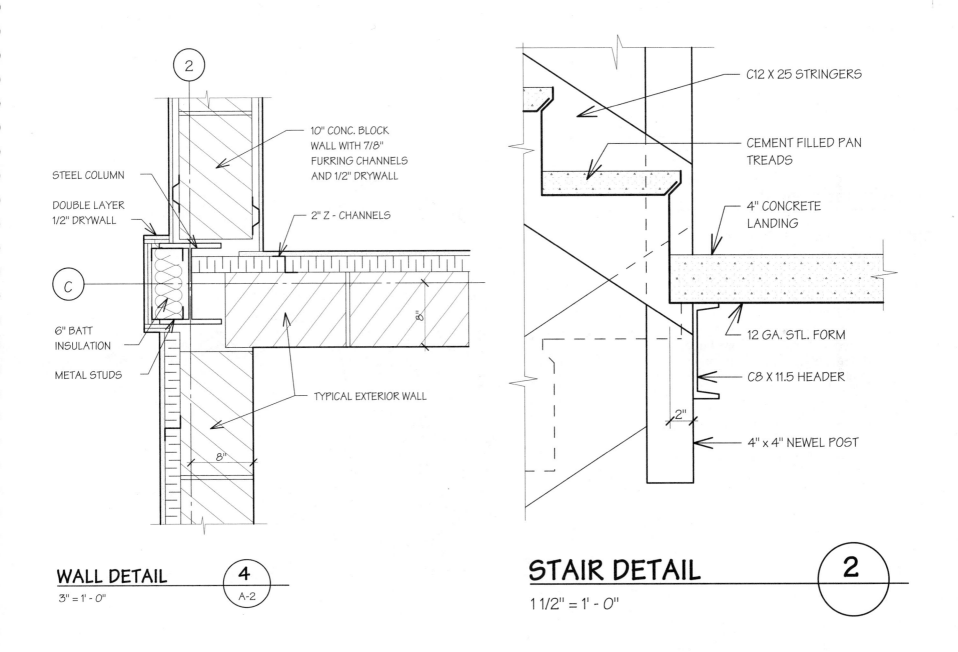

**WALL DETAIL**

3" = 1' - 0"

④
A-2

**STAIR DETAIL**

1 1/2" = 1' - 0"

②

Labels in wall detail:
- STEEL COLUMN
- DOUBLE LAYER 1/2" DRYWALL
- 6" BATT INSULATION
- METAL STUDS
- 10" CONC. BLOCK WALL WITH 7/8" FURRING CHANNELS AND 1/2" DRYWALL
- 2" Z - CHANNELS
- TYPICAL EXTERIOR WALL
- 8"
- ②
- C

Labels in stair detail:
- C12 X 25 STRINGERS
- CEMENT FILLED PAN TREADS
- 4" CONCRETE LANDING
- 12 GA. STL. FORM
- C8 X 11.5 HEADER
- 4" x 4" NEWEL POST
- 2"

# KEY TERMS

masonry opening
threshold
weather-stripping
undercut
louver
label

# PRACTICE EXERCISE

For the practice exercise project begun in the last two chapters, complete the following:

- floor plans
- room finish schedule
- door schedule
- 2 building sections
- 2 wall sections
- exterior elevations for one building
- interior elevations for one room
- 3 details

Use the sketches on the following pages as a guide but use references to find any information that is missing.

Continue with the same sheet layout you have used previously. Cross reference all drawings in the set.

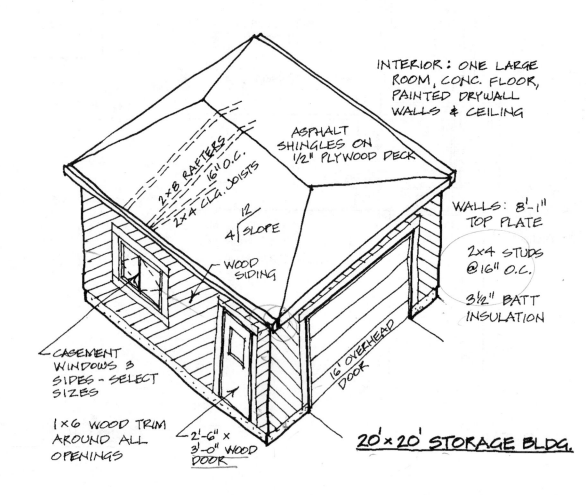

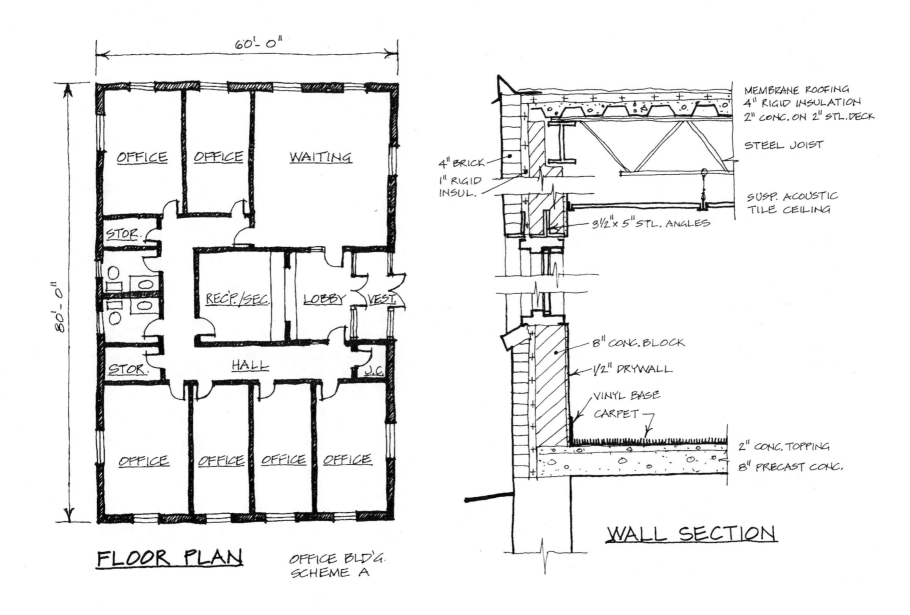

60'- 0"

80'- 0"

OFFICE   OFFICE   WAITING

STOR.

RECP./SEC.   LOBBY   VEST.

STOR.   HALL   J.C.

OFFICE   OFFICE   OFFICE   OFFICE

FLOOR PLAN

OFFICE BLD'G.
SCHEME A

MEMBRANE ROOFING
4" RIGID INSULATION
2" CONC. ON 2" STL. DECK

STEEL JOIST

4" BRICK

1" RIGID INSUL.

SUSP. ACOUSTIC
TILE CEILING

3½" x 5" STL. ANGLES

8" CONC. BLOCK

½" DRYWALL

VINYL BASE
CARPET

2" CONC. TOPPING
8" PRECAST CONC.

WALL SECTION

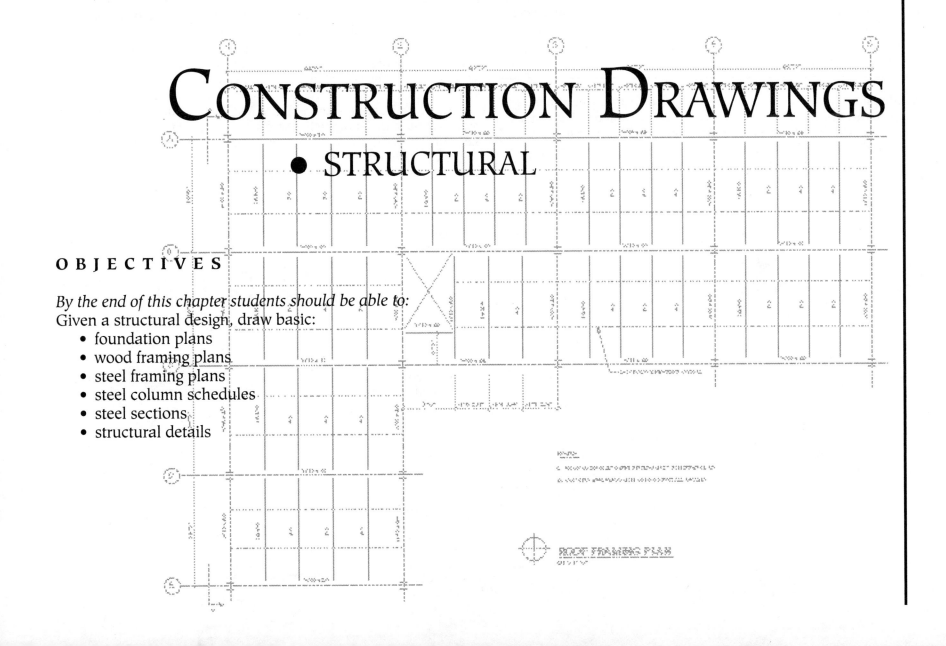

# CONSTRUCTION DRAWINGS

## • STRUCTURAL

### OBJECTIVES

*By the end of this chapter students should be able to:*
Given a structural design, draw basic:
- foundation plans
- wood framing plans
- steel framing plans
- steel column schedules
- steel sections
- structural details

Structural drawings are organized in the same manner they are designed. The **sub-structure** (the part below the surface) is one entity and is most commonly referred to as the **foundation**. The **super-structure** (the part above the surface) is the other entity and is most commonly referred to as the **framing**. Additionally, as in other disciplines, schedules and details are required to completely describe the structure.

# FOUNDATIONS

Whether the structure above ground is wood, steel, or concrete, foundation plans are essentially the same because they are almost always made of concrete. The foundation can be cast-in-place or in the form of masonry units (concrete blocks).

Foundations are designed to support two loading conditions: uniform loads and concentrated loads. Uniform loads are those produced by walls and floors; concentrated loads are those produced by columns.

In the case of uniform loads, a continuous foundation wall must support the weight. For a concentrated load, a **pier** supports the weight. A **pilaster** is a pier combined with a foundation wall.

In most cases the load is too great to be transferred to the soil directly, so a **footing** is used to distribute the load over a greater area. The weaker the soil, the wider the footing needs to be, as determined by building codes.

Additionally, **reinforcing** is added to all parts of the foundation system to increase its strength. These steel bars are laid out in grids or bent into shapes to strengthen specific areas or to help connect one part of the foundation to another. In a floor slab, wires welded into a mesh or grid pattern are used to hold this relatively thin piece of concrete together.

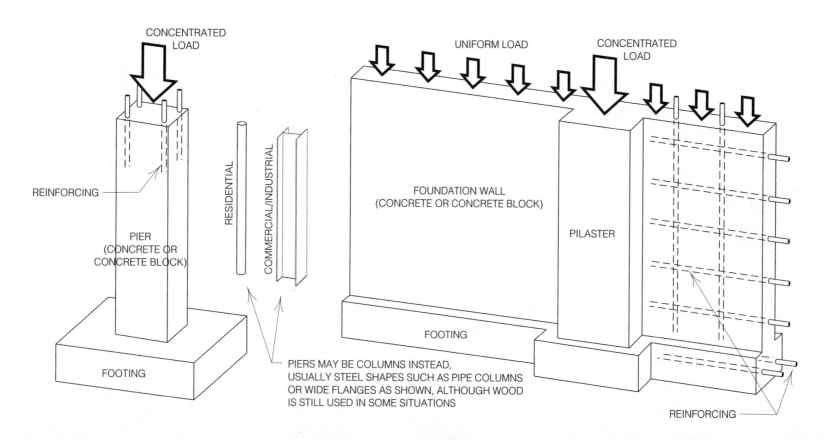

CONCENTRATED LOAD

REINFORCING

PIER (CONCRETE OR CONCRETE BLOCK)

FOOTING

RESIDENTIAL

COMMERCIAL/INDUSTRIAL

PIERS MAY BE COLUMNS INSTEAD, USUALLY STEEL SHAPES SUCH AS PIPE COLUMNS OR WIDE FLANGES AS SHOWN, ALTHOUGH WOOD IS STILL USED IN SOME SITUATIONS

UNIFORM LOAD

CONCENTRATED LOAD

FOUNDATION WALL (CONCRETE OR CONCRETE BLOCK)

PILASTER

FOOTING

REINFORCING

It must be understood that foundations are built several different ways. If the ground floor is a concrete slab on grade, then the foundation is completely underground. The only excavation required is for the foundation walls and footings themselves.

If the ground floor is not supported by the earth itself, but rather is supported by piers and foundation walls, one of two conditions exists. The first condition is one in which a small space is excavated so that work and service on the floor can occur. It is not high enough for a person to stand in; therefore it is called a ***crawl space***. The depth is usually a function of the ***frost line***, the depth to which the moisture in the ground freezes. Again, building codes determine this depth.

If the foundation does not extend below the frost line, then the moisture or water present in the soil can actually push a building up when the water freezes and becomes ice. In colder climates where foundations must go fairly deep, the decision may be made to extend the crawl space down so that a full story results. What we then have is a basement where the foundation plan and basement plan become one.

There may, of course, be the situation where a full basement is desired by the client regardless of the climate. It is not uncommon for skyscrapers to have several stories below grade for parking, mechanical systems, storage, etc.

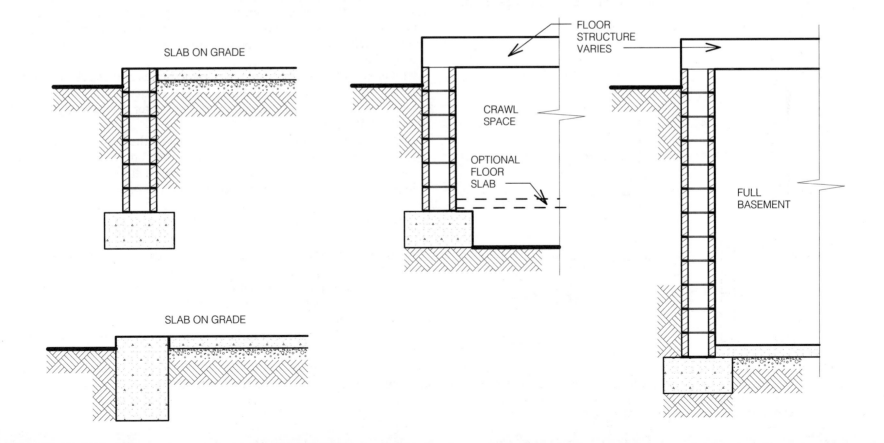

# FOUNDATION PLANS

## Foundation Wall

The first step in producing a foundation plan is to outline the foundation walls. Drafters must be familiar with the detail of how the exterior wall meets the foundation wall. In the majority of cases, the outside faces of the two walls are aligned. This always should be checked rather than assumed however.

The outline of the foundation may be adapted from the floor plan, which is usually fairly well defined at this stage. If the outside face is the same, the outside face of the floor plan can be traced. Then it is simple enough to put the wall thickness of the foundation wall inside of that. For this reason the foundation plan is usually drawn to the same scale as the first floor plan.

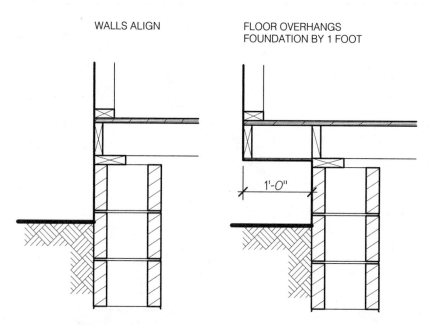

WALLS ALIGN                FLOOR OVERHANGS
                          FOUNDATION BY 1 FOOT

1'-0"

*The outside face of the foundation wall is commonly aligned with the outside face of the first floor wall. In some structures, particularly wood frame buildings, the first floor wall may extend beyond the foundation wall. The drafter must be aware of any such difference between the plans and document them properly.*

## Piers

Piers are the next element in a foundation plan. Since piers support loads from columns directly above them, they are centered under the columns. Once again, when a pier is located within an outside wall so that the pier and foundation wall are combined, it is called a pilaster.

Since piers and foundation walls are in section, they should be drawn as thick lines. When engineers draw foundation plans, they tend not to indicate the material of the wall. When architects draw foundation plans, they do indicate the material.

## Structural Grid

If there are columns in the super-structure, they are almost always arranged in a uniform pattern or grid. The center lines of the columns (and therefore the piers and pilasters) must be indicated with continuous center lines that extend beyond the building line on the top or bottom and to the left. These lines will be used to dimension the piers and pilasters later.

Reference tags or circles are placed at the ends of these center lines. They are drawn about 3/8" in diameter and are lettered from A to Z vertically and from 1 to "n" horizontally. The center lines and tags should be drawn with thin lines.

The vertical tags are almost always found on the left. The horizontal tags, however, are found as frequently on the bottom as they are on top. In either case, the lettering and numbering begins at the vertex of the tag strings, i.e., "A1" would be the first intersection of tags.

## Footings

The next objects to be drawn on the plan are footings. These are objects beyond the cutting plane and are drawn as solid, medium lines. If there is a floor slab, as in the case of a crawl space and/or basement, then the inside edge of the footing is covered with the slab and should be shown as a hidden line.

The outer edge of the footing is covered with earth, so technically it too should be shown as a hidden line. However, for the sake of clarity it is assumed that the slab and earth are not there. The accepted method for drawing footings is with solid lines, as shown on the facing page.

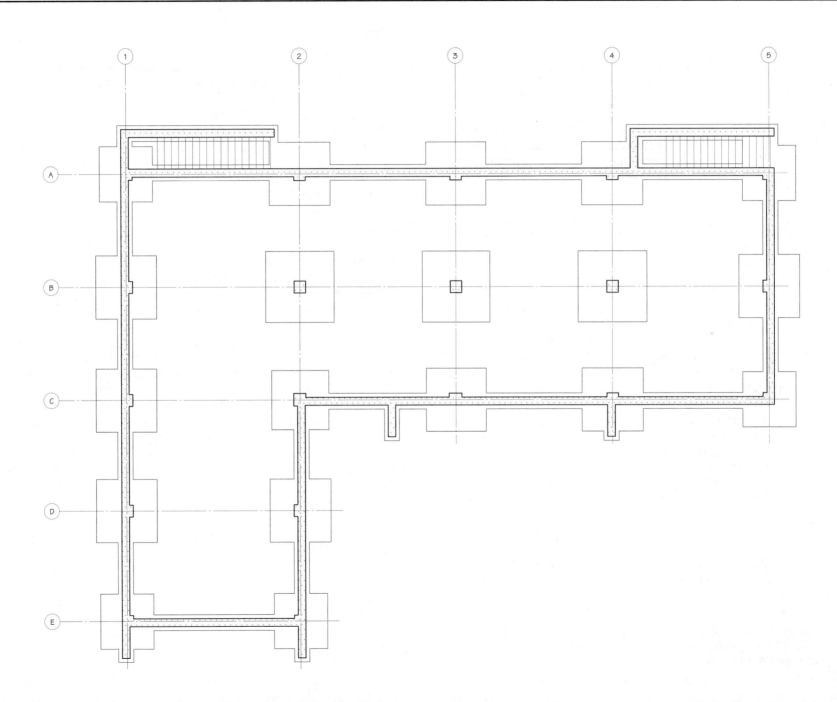

## Slabs

The final part of the foundation that must be indicated is the floor slab, should there be one. In a plan view, the edge of the slab is defined by the foundation wall so there is really nothing to draw. The slab is indicated with a note.

The note should be centered and lettered with medium sized lettering, about 1/8", and then outlined with a border so that it stands out. Typically the thickness, material, and name are given followed by the reinforcing it should have within it. It is also common to include the finished floor elevation.

At this point, the entire foundation is represented. Now the dimensioning, notes, and reference symbols can be added.

## Dimensioning

If the foundation is made of concrete masonry units, it is dimensioned using overall and offset strings. If the foundation is a basement, then it should be treated as an architectural floor plan.

Cast-in-place concrete foundations are dimensioned using overall and offset strings except when they support a skeletal structural system, such as in a larger building. In this situation it is much easier to dimension the grid and then locate the foundation wall and footings around it.

As our example on the facing page illustrates, the grid has already been established and tagged. The first step is to use the grid center lines as the extension lines and dimension them. Dots are the most common terminators used when dimensioning to center lines. However, it is not incorrect to use any of the other terminators.

The walls, piers, and footings all relate to the grid. The piers and their footings are centered where two grid center lines intersect, so they do not need to be dimensioned. The foundation walls and their footings do, however, need to be dimensioned in relation to the grid lines they straddle.

These few dimensions can either be placed on the plan itself or incorporated into the details of each pilaster. It is helpful to place dimensions on both the plan and the details.

The dimensions of piers and footings may be given in schedules. This is a much more efficient way of providing the information and certainly eliminates what would be a tedious job for the drafter. We will discuss schedules later in this chapter.

## Notes

Once the foundation has been dimensioned, any notes that are needed can be lettered. For foundation plans, notes are used to point out any exceptions to what has already been drawn since every object has been drawn, noted, and dimensioned at this point.

The notes that tend to be needed are those related to special reinforcing, or special conditions or instructions to the contractor. These may be done with small lettering and leaders and terminators.

## Reference Symbols

The last step of the foundation plan is to add all of the reference symbols. The drafter should place a north arrow, letter the title and scale to standards, and draw section and detail marks.

The last symbol that must be added is for identifying footings. This symbol needs to be added next to each footing to provide information about it and to link it to the footing schedule, which will be discussed next.

Two pieces of information need to be given on the foundation plan: a footing identification letter/number (F1, F2, etc.) and the elevation (height) of the footing. It is not necessary to give each footing its own identification because many of them will be exactly the same. Instead, a specific footing design (size, reinforcing, etc.) is identified and assigned to each applicable footing. This footing identification is the link to the footing schedule, which provides all the design information.

The footing elevation is generally measured to the top. This makes it the same elevation as the bottom of the pier. Alternately, the bottom of the footing is given. A typical pier/footing symbol is shown below. It should be about a 1/2" square.

| F2 | FOOTING IDENTIFICATION |
| 84.13 | TOP OF FOOTING ELEVATION |

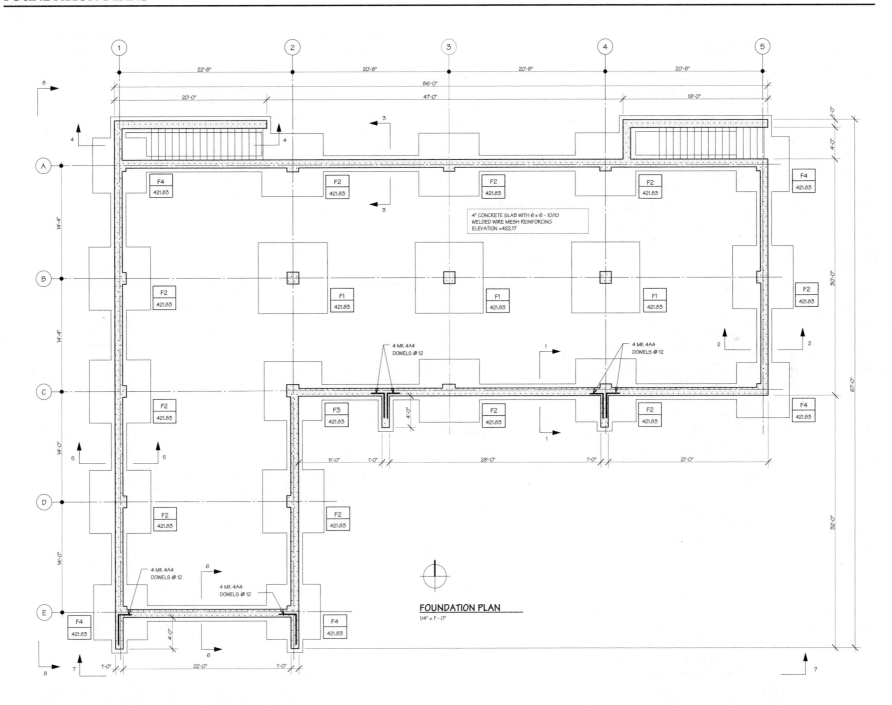

4" CONCRETE SLAB WITH 6 x 6 - 10/10
WELDED WIRE MESH REINFORCING
ELEVATION +422.17

**FOUNDATION PLAN**
1/4" = 1' - 0"

# FOUNDATION SCHEDULES

Schedules are usually not needed for residential or small framed commercial buildings because their foundations are not very complicated. For larger structures, schedules are an effective way of providing detailed information about the foundation. The two essential schedules are footing schedules and pier schedules.

## Footing Schedule

The first column in a footing schedule is the footing mark. A **mark** is nothing more than the identification letter/number given to an object. Since several footings may be given the same mark, the second column in the schedule gives the quantity. Next the size is given in length, width, and depth. Notice that footings are almost always square and that the depth is the last figure given.

The final two columns relate to reinforcing. Reinforcing is the network of steel bars encased in concrete to help hold the footing together and to resist some of the forces acting on it. There are two types of reinforcing for footings: straight bars for the footing itself and "L" shaped bars, called **dowels**, to connect the footing to the pier above it.

The fourth column in the schedule lists "REINF. 1/2 E.W." For footing F2 this is 20-#7 x 7-6. This means there are 20 reinforcing bars, their size is #7, and they are 7'-6" long. Bar numbers are their diameter measured in 1/8" increments; thus a #7 bar is 7/8"Ø, a #6 is 6/8" or 3/4"Ø, etc. The column heading says 1/2 E.W. (each way), meaning that one half of the 20 bars (10) extend horizontally one way and the other 10 extend perpendicular to the first 10.

Again for footing F2, the final column shows that there are 8 dowels of mark 8A35, abbreviated "MK8A35." This mark links to another schedule, "bending details," which gives the size and dimensions of this and other bent bars.

## Pier Schedule

Pier schedules are similar to footing schedules. The first column in the pier schedule lists the piers. These designations are based on the structural grid, "A1" being the pier in the upper left of the plan, etc.

The next two columns in a pier schedule list the quantity and size of the piers in inches. This is followed by the vertical reinforcing: number of bars, size, and length.

The final column is entitled "ties." **Ties** are reinforcing bars wrapped around vertical bars to hold them in place. They are horizontal, square-shaped reinforcing bars. The number required and their mark are listed in this last column. Their spacing is noted in the column heading. Like dowels in the footing schedule, specifics of this reinforcing may be found in the bending schedule.

The configuration of footing, pier, and their reinforcing is illustrated on the facing page.

| FOOTING SCHEDULE | | | | |
|------|-------|---------------|----------------|-------------|
| MARK | QUAN. | SIZE | REINF. 1/2 E.W. | DOWELS |
| F1 | 6 | 9-0 x 9-0 x 1-11 | 20 - #8 x 8-6 | 8 - MK 9A34 |
| F2 | 1 | 8-0 x 8-0 x 1-10 | 20 - #7 x 7-6 | 8 - MK 8A35 |
| F3 | 4 | 7-6 x 7-6 x 1-8 | 18 - #7 x 7-0 | 6 - MK 8A37 |
| F3A | 1 | 7-6 x 7-6 x 1-8 | 18 - #7 x 7-0 | 8 - MK 8A37 |
| F4 | 5 | 7-0 x 7-0 x 1-7 | 16 - #7 x 6-6 | 6 - MK 8A38 |
| F5 | 3 | 6-6 x 6-6 x 1-6 | 18 - #6 x 6-0 | 6 - MK 8A39 |

| PIER SCHEDULE | | | | |
|------|-------|---------|----------------|--------------|
| PIER | QUAN. | SIZE | VERTICAL | TIES @ 12" |
| B2, B3, B4 C2, C3, C4 | 6 | 16 x 16 | 8 - #9 x 10-2 | 10 - MK 3A1 |
| A1 THRU A5 B1, B5, C1, C5 D2, D5 | 11 | 16 x 16 | 6 - #8 x 10-2 | 10 - MK 3A1 |
| D1 | 1 | 16 x 16 | 6 - #8 x 13-2 | 13 - MK 3A1 |
| D3, D4 | 2 | 16 x 16 | 8 - #8 x 12-2 | 12 - MK 3A1 |

# FOUNDATION DETAILS

As mentioned in the foundation plan description, it is common to enlarge parts of the plan to form details. The most common details are those of the piers and pilasters (sometimes called exterior piers), which when enlarged become easier to dimension and show reinforcing.

The procedure for foundation details is the same as that for details of the parent drawing. The object is outlined, object lines are drawn, materials are indicated, dimensions and notes are added, and finally title and scale are added.

Below are shown typical details at their original size of 1/2" = 1' - 0". They are normally placed on the same sheet as the foundation plan itself. This makes reading the drawings much easier.

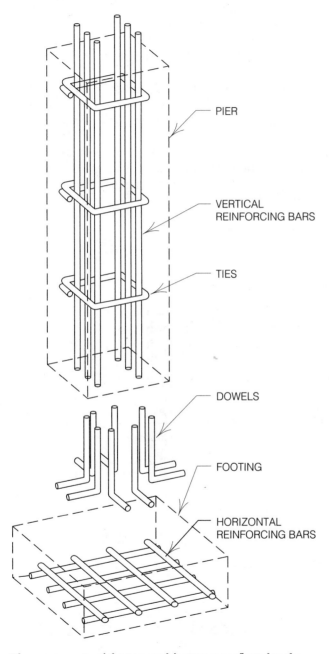

PIER

VERTICAL
REINFORCING BARS

TIES

DOWELS

FOOTING

HORIZONTAL
REINFORCING BARS

*The components of the pier and footing are reflected in the schedules at left and the details at right.*

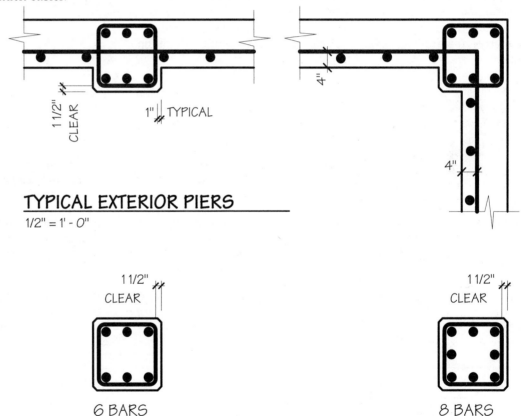

1 1/2"
CLEAR

1" TYPICAL

4"

4"

## TYPICAL EXTERIOR PIERS

1/2" = 1' - 0"

1 1/2"
CLEAR

1 1/2"
CLEAR

6 BARS

8 BARS

## TYPICAL INTERIOR PIERS

1/2" = 1' - 0"

# FOUNDATION SECTIONS

Sections are in essence vertical details. They are enlarged drawings of cuts through the wall at various locations. Their size makes it easier to show dimensions and reinforcing locations.

Typical sections are shown below. There usually is a typical pier and footing section, and sections wherever the wall, footing, or reinforcing changes. In each case the procedure is the same because the main function of the section is to show reinforcing and give dimensions.

The section shows the key elements of the foundation: footing, wall or pier, slab, and reinforcing. The notes state the number, size, and spacing of reinforcing or their marks for cross referencing to another location on the drawings. Notes usually fit easily onto one side of the section.

A new type of terminator is needed for identifying reinforcing. When a bar runs perpendicular to the picture plane, it shows as a small blackened circle. This is tough to point to exactly with an arrow, so a circle or loop is used to identify it.

Bent reinforcing bars like the dowels and ties already shown are noted and keyed to the bending schedule. A key dimension for all reinforcing is its clear space from the edge of the concrete. This is needed for protection.

Dimensioning of the concrete or block itself need not be added because it is documented elsewhere. The foundation plan gives length and width, and the elevations give heights. Thus dimensions in a foundation section should be limited to the minor variations that are best seen in section.

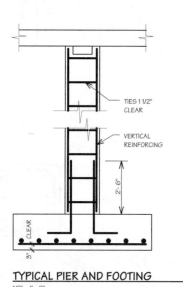

**TYPICAL PIER AND FOOTING**
1/2" = 1' - 0"

*On the left is a typical pier and footing section. Notice that because this is typical, it applies to the majority if not all of the piers and footings on the project. This means that the dimensions vary, so the configuration is the main purpose of the section. The only constant for all footing types is the height to which the dowels are located and the clear space between the face of the concrete and the reinforcing bars.*

*On the right is a section at a specific location as the title defines. Here the exact reinforcing can be shown and labeled. Also important is the clear dimension from the face of the concrete to the edge of the reinforcing. The notes, though abbreviated, are efficient. The main vertical bars in the wall, for example, are noted as "66-#4 x 10' - 0 @ 12 O.C. OMIT AT PIERS." That means that there are 66 #4 bars that are 10' long. They are spaced 12" apart, and they stop at each pier.*

*The dowels are noted almost the same way. The only difference is a mark designation rather than a bar number.*

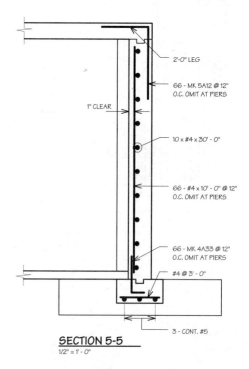

**SECTION 5-5**
1/2" = 1' - 0"

# FOUNDATION ELEVATIONS

As shown below, the wall and footing are outlined with medium object lines since nothing is in section. The drafter may want to vary line weights somewhat to give some illusion of depth.

If a pilaster occurs on the inside face of the wall, it is shown with a hidden line, and the grid line associated with the pilaster is drawn complete with tag and mark.

If openings occur, they are drawn and marked with an "x" and then dimensioned in both directions. Any other indentations such as brick ledges or recesses in the wall are also dimensioned where they occur.

Two key notations are the elevation of the top of the wall and the elevation of the top of the footing. Each is entered next to an extension line at the height it defines. The elevation number (text) is of medium height lettering such as 1/8".

The most important information shown on elevations is the placement of reinforcing. In the example below, extension and dimension lines are placed to define the area in which the reinforcing fits, both vertically and horizontally. A notation along the dimension line gives the reinforcing information. Individual reinforcing bars are noted and indicated with leaders and terminators as needed.

Notes of special conditions or instructions are placed where needed and terminators are used to locate them. These general notes can be of smaller size such as 1/8" or 3/32".

The drawing is completed with title, scale, and reference symbols such as section or detail marks. Grid lines, with their corresponding tags, are also added. Typically, elevations can be of the same scale as the plan. If they are complicated, a larger scale may be needed. It may also be necessary to draw two elevations of the same wall, one to show the concrete work and the other to show the reinforcing work.

This completes the basic drawings for foundation documentation. It is desirable to place all of these drawings together on one sheet as they are highly interdependent.

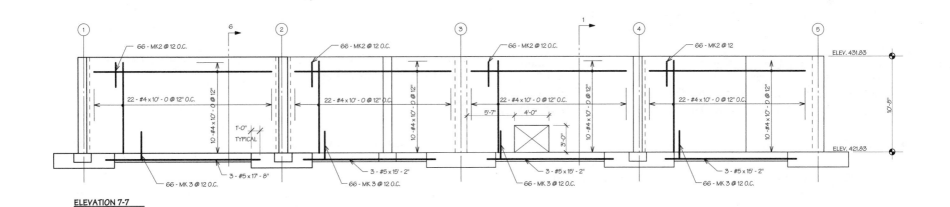

ELEVATION 7-7
1/4" = 1' - 0"

# FRAMING

Framing drawings are required to show the structure of a building above the ground. Generally, a building's super-structure is composed of one or more of three materials: wood, steel, and concrete.

# WOOD FRAMING PLANS

## Structural Members

A framing plan is a view looking down and is the same as floor and roof plans except that the floor or roof itself has been removed to expose the structural members that support it.

Wood framing plans are fairly simple to produce because the drafter just draws the members as they are seen and then labels what they are. For light framing the members are usually 2" wide nominally (1-1/2" actual) and are drawn as long rectangles 1-1/2" wide. At small scale these thin members are usually drawn as single lines.

Heavy timber framing members are those that are over 4" in their smallest dimension. They may either be solid pieces of wood or composites made from smaller pieces.

All members in a wood framing plan are objects and thus are drawn as continuous, solid lines. Since the members are not in section the lines should be medium weight.

The drawing on the facing page is a basic floor framing plan using 2 x 10 joists. **Joists** are regularly spaced small beams that support a floor or roof. In this case the joists are spaced 16" apart, measured from the center of one to the center of the next one. This is noted as 16" **on center** (o.c.).

## Openings

The drawing on the facing page contains a cut out in the center of the plan where there are no joists. It is marked with an "x" meaning it is open. Openings such as this are made for chimneys, elevators, shafts, or in this case, stairs.

Where an opening occurs, the joists adjacent to the opening are interrupted and deprived of their support at one end. To support these joists, the drafter needs to place a beam perpendicular to them. As is done here, the perpendicular beam is usually two members the same size as the joist. This type of beam is called a **header**.

Once the joists have been supported, the headers now need to be. The headers are carrying the weight from the shortened joists out to the first full length joists at either side of the opening. Since a single joist at that location is not capable of carrying its own weight and that of the headers, it too must be made stronger. Again this is usually done using two or more of the same size member depending on the actual load. In this case a double 2 x 10 is needed on one side. The other side has a glue laminated beam.

## Support Beams

Joists are supported by the foundation wall around the perimeter of the building, but in most cases joists are too weak to span the entire width. Economically it is better to add a beam for support at mid-span (or at one third or one fourth points) than to use bigger joists.

When a beam is used to reduce the joist span, then the joists do not need to be continuous. This is the case on the facing page where there are two sets of joists on the left. They are lapped where they meet above the steel beam and are nailed together at that point.

In our example there are two beams. A steel beam supports the joists mid-way on the left, and a glue laminated beam supports one end of the joists on the right. A glue-lam beam used here would be unusual but is included to introduce the concept.

**Glue laminated** beams, as the name suggests, are layers of thinner pieces of wood (about 1-1/2" thick) glued and pressed together to make larger, stronger beams. They are more readily available and stronger than solid timber beams of the same cross section area.

## Notes

The technique for noting members is to give their size directly adjacent to the member. If there are multiple members, as is the case for the headers, then the number is also noted as shown.

Since this technique can get a little tedious for the joists, an easier convention is used. The size and spacing of the joists is printed perpendicular to them and then an arrow is projected in each direction to the length the joists span. This is a quick and easy way to convey the information.

Heavy timber framing works exactly the same way. Because the spacing tends to be much wider than joists, the short-cut convention is not necessary.

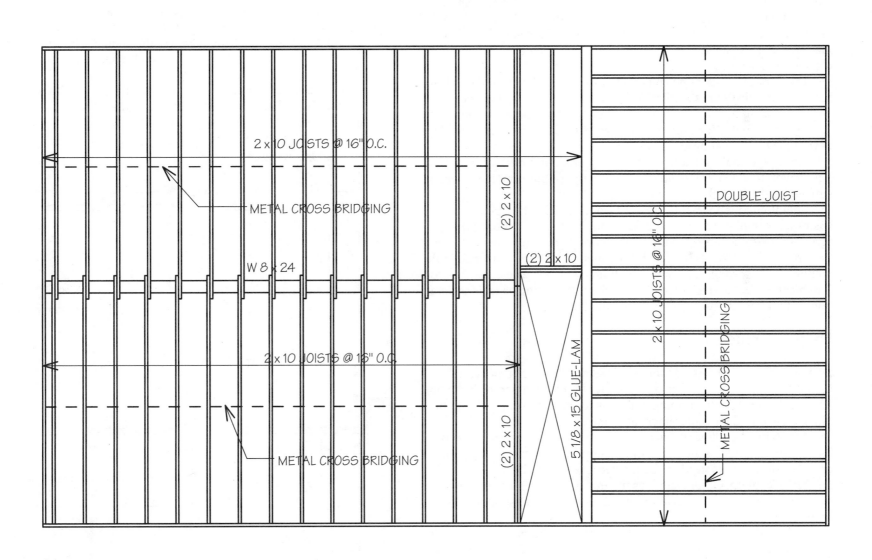

2 x 10 JOISTS @ 16" O.C.

METAL CROSS BRIDGING

W 8 x 24

2 x 10 JOISTS @ 16" O.C.

METAL CROSS BRIDGING

(2) 2 x 10

(2) 2 x 10

(2) 2 x 10

5 1/8 x 15 GLUE-LAM

2 x 10 JOISTS @ 16" O.C.

DOUBLE JOIST

METAL CROSS BRIDGING

## Miscellaneous Members

Another member that must be drawn is **cross bracing** or **cross bridging**. These wood or metal pieces help hold joists upright and also help to spread the load among joists. They are frequently drawn as single, hidden lines even though they are visible. They are noted the same as other members but usually are not dimensioned since they are assumed to be centered in a span.

The floor of wood frame construction is built in layers and needs to be noted. A **sub-floor** is applied first to connect the joists together and to cause the resulting platform to work as a unit. Plywood is the most common material. **Plywood** is a sheet good made from several thin layers of wood with alternating grain direction.

Since plywood is a relatively rough material, an **underlayment** is the next material added. It has no structural purpose but is merely a smooth surface upon which the last layer can be added.

Common underlayments are luan wood, composition board, or a smoother type of plywood. It is also common practice to utilize thicker plywood sheets as a combination sub-flooring and underlayment.

The final material is the finish floor material, which is defined in the architectural drawings. However, the framing plan is where the sub-floor (often called decking) is noted. As shown on the facing page an isolated note with a border around it spells out the flooring.

## Dimensioning

Dimensioning of light wood framing plans need not be complicated. Member spacing is certainly important but that is included with the note for each member for light framing. Beams for heavy framing should be dimensioned to their center lines or to one edge.

The only other dimensions are those that locate a change from the basic overall framing scheme. In our example on the facing page, the opening for the stairs needs to be located. It needs to be a precise size at a specific location. Other examples include a change in the size of members, an overhang, or the location of a double joist that supports an additional load.

For wood framing we give the carpenter a good amount of flexibility in laying out the work. Use dimensions when a structural member must be placed at a specific location.

The framing plan is completed by adding title, scale, and reference symbols. The completed framing plan is shown on the facing page.

# WOOD SECTIONS, ELEVATIONS, AND DETAILS

Structural sections, elevations, and details are rarely needed for light wood construction. They are a little more common for heavy timber. This is because wood structures are usually fairly simple and can be shown within the architectural drawings where all the materials can be shown together.

Below are structural details for the plan at right to help you understand some of the basics of wood frame construction.

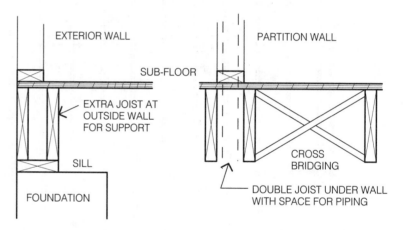

VIEW PARALLEL TO JOISTS

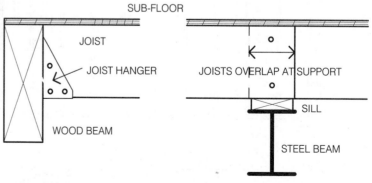

VIEW PERPENDICULAR TO JOISTS

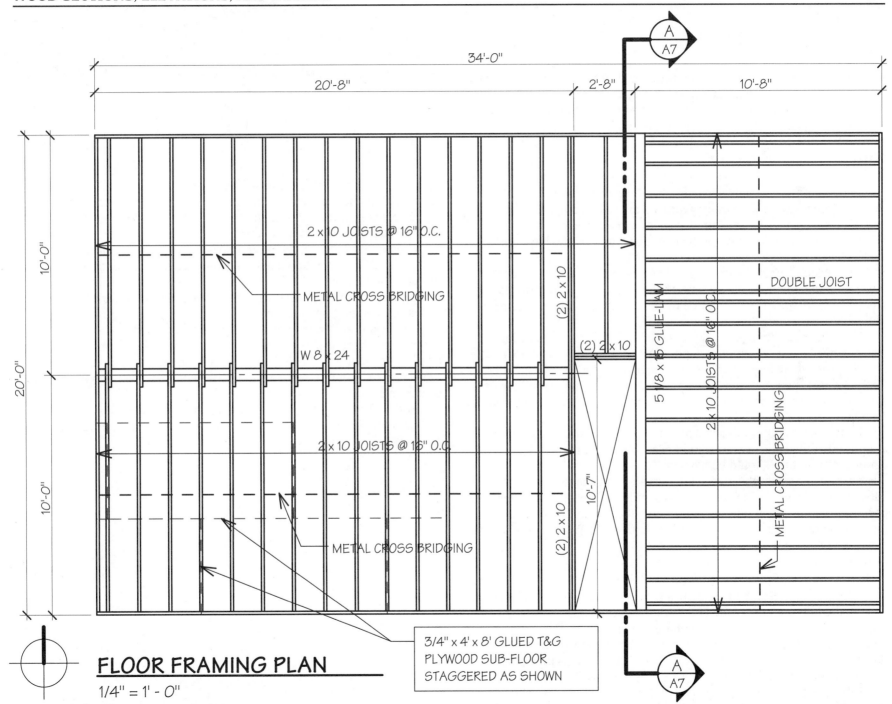

34'-0"

20'-8"

2'-8"

10'-8"

10'-0"

20'-0"

10'-0"

10'-7"

A
A7

2 x 10 JOISTS @ 16" O.C.

METAL CROSS BRIDGING

(2) 2 x 10

W 8 x 24

(2) 2 x 10

2 x 10 JOISTS @ 16" O.C.

(2) 2 x 10

METAL CROSS BRIDGING

5 1/8 x 15 GLUE-LAM

2 x 10 JOISTS @ 16" O.C.

DOUBLE JOIST

METAL CROSS BRIDGING

3/4" x 4' x 8' GLUED T&G
PLYWOOD SUB-FLOOR
STAGGERED AS SHOWN

A
A7

## FLOOR FRAMING PLAN

1/4" = 1' - 0"

# STEEL FRAMING PLANS

Conceptually a steel framing system is drawn exactly the same as a wood system. The main difference is that steel structures are usually designed on a grid pattern. In fact, the first step in producing a steel framing plan is to lay out the grid.

## Grid

A grid is the most economical way to utilize the benefits of steel. What the grid does is divide the building into a matrix of rectangles. A **bay** is one of these rectangles — the area formed by the arrangement of four columns. It is the center lines of the columns that form the bays and create the grid.

On the facing page, the main grid has four bays in the east-west direction and either four or two bays in the north-south direction. There are ten bays in all. The grid is made by the center lines of the columns; thus they are thin lines.

The grid lines should extend on the top or bottom and to the left to some degree. The exact amount is not important. The goal is to extend the lines beyond where any other lines or dimensions may be placed later. Basically these lines are to be used as extension lines for overall dimensions and should follow the conventions established in the dimensioning section of this book.

The grid must now be labeled for easy identification. Circles or tags are placed at the ends of all the grid lines. The size of the tags should be about 3/8", and they should be aligned.

Starting at the left, the horizontal tags should be labeled with numbers beginning with 1. The vertical tags should be labeled with letters beginning with A. Start lettering at the top if the horizontal tags are at the top; start lettering from the bottom if the horizontal tags are at the bottom.

## Columns

With the grid complete the next step is to draw the columns that occur at each of the intersections of the grid lines. Most steel framing systems utilize wide flange steel columns that are shaped like an "H." They may be oriented or rotated in any direction but their center will usually be at the intersection of the grid lines. They should be drawn as thick object lines since they are in section.

As we will see later, the columns are defined in a schedule so there is no need to label them on the plan. Actually they have already been labeled by way of the grid coordinates they fall on. The column in the lower left hand corner, for example, is known as E1, and the one at the upper right is A5.

## Beams

The next step for the framing plan is to draw the beams located above the columns just drawn. Structural steel frames generally have beams that connect all of the columns to each other. The four corner columns have only two beams attached to them, the other exterior columns have three, and the interior columns have four.

Unlike wood beams, the drafter does not draw the width of structural steel beams. They are drawn and dimensioned to their center lines because they are assembled that way in the field. Thus, beams are drawn as single lines representing their center. The lines are rendered as solid object lines of medium line weight.

The beams are drawn between the columns but stop short of touching them. A small gap is shown even though in reality the beams are connected to the columns. This makes the drawing easier to read, although drawing the structural members in this way almost covers up the grid of center lines that was drawn initially.

At this point the main framework has been completed. All that remains is to draw the intermediate steel beams or joists between the main ones. These are drawn the same as the other beams.

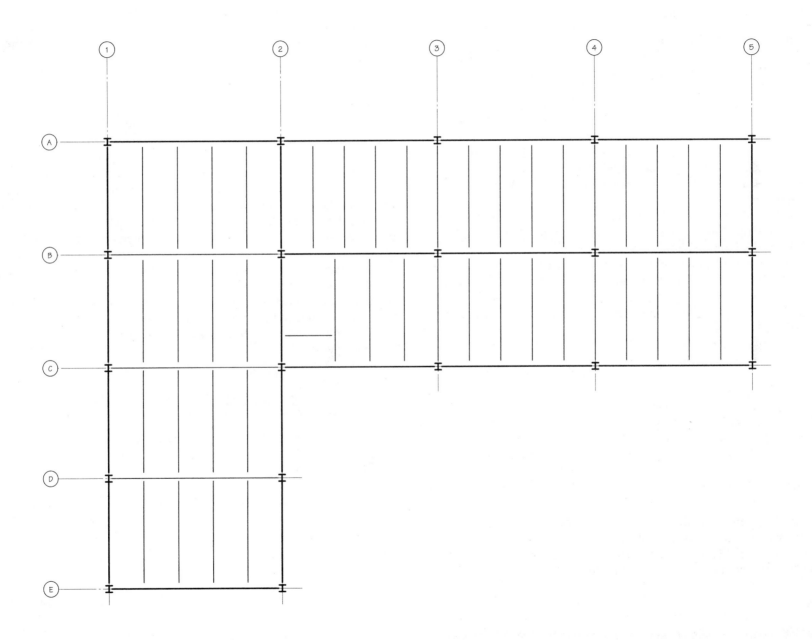

## Notes

Now all the beams and joists must be labeled using standard beam and joist designations. Labels for beams running left to right are placed slightly above the line. Labels for beams running up and down are placed to the left of the line in a similar fashion. The beam or joist label is centered on the member and should be in small lettering, 3/32" to 1/8" maximum.

You will notice that several members are labeled "do." This is an abbreviation for the word "ditto," which means "the same as above." Thus all the structural members labeled that way are the same as the original one labeled above it. This convention saves the manual drafter a lot of lettering time.

The height of the steel beams is usually recorded on other drawings such as a schedule. Therefore there is no need to indicate how high the beams are on the framing plan.

Occasionally a beam needs to be higher or lower than the beams around it. This is usually done to accommodate a special floor. In this situation a height variation is added to the note for that particular beam. A beam set two inches lower than the established top of steel would be noted "W 8 x 10  (-2")."

## Miscellaneous Members

Steel framing, especially steel joists, needs bridging just as wood framing does. In the case of steel, bridging is usually a round steel rod or a steel angle welded to the top and bottom of each member. Bridging is shown on the framing plan as a single hidden line, and its structural shape is noted. The most common structural shapes are shown at right.

## Openings

There may be openings in the floor or roof that need to be framed with larger beams. This is done the same way as with wood, although the larger beams are not called headers. These framing members are noted and an "x" is drawn in the opening with thin lines.

## Deck

The only thing that remains to complete the framing plan is to note what the floor or roof is to be made of. In our example on the facing page, note

number 1 of the general notes clearly states the deck and slab to be used. Another method is to locate the note in the center of the plan and outline it. This was the method used earlier in the wood framing plan. Either method is acceptable.

## Dimensioning

Most of the steel frame dimensioning is accomplished when the grid is dimensioned. As was done on the foundation plan, the tagged ends of the grid are used as extension lines and each bay is dimensioned.

These dimensions locate all the columns and the main beams since they are centered on the columns. Now all that needs to be dimensioned are the joists.

Usually the joists are spaced evenly across a bay. Since many of the bays are the same size, it is acceptable to dimension only one bay and mark it as "typical," meaning it applies to all other bays that are the same size. If a bay differs from the norm then it is dimensioned individually.

Like wood framing, the only other dimensions needed on steel framing plans are those to locate critical members, size openings, or to indicate a change to the basic structural scheme.

The drawing is completed by adding the title, scale, north arrow, and reference symbols.

| SHAPE | DESCRIPTION | DESIGNATION | EXAMPLE |
|-------|-------------|-------------|---------|
|  H | WIDE FLANGE | W, M | W8 x 24 |
|  I | STANDARD | S | S12 x 35 |
| T | TEE | WT, ST | WT18 x 115 |
|  LL | ANGLE | L | L5 x 3 x 1/4 |
| [ | CHANNEL | C, MC | MC10 x 28.3 |
|  ʐ | ZEE | Z | Z4 x 8.2 |

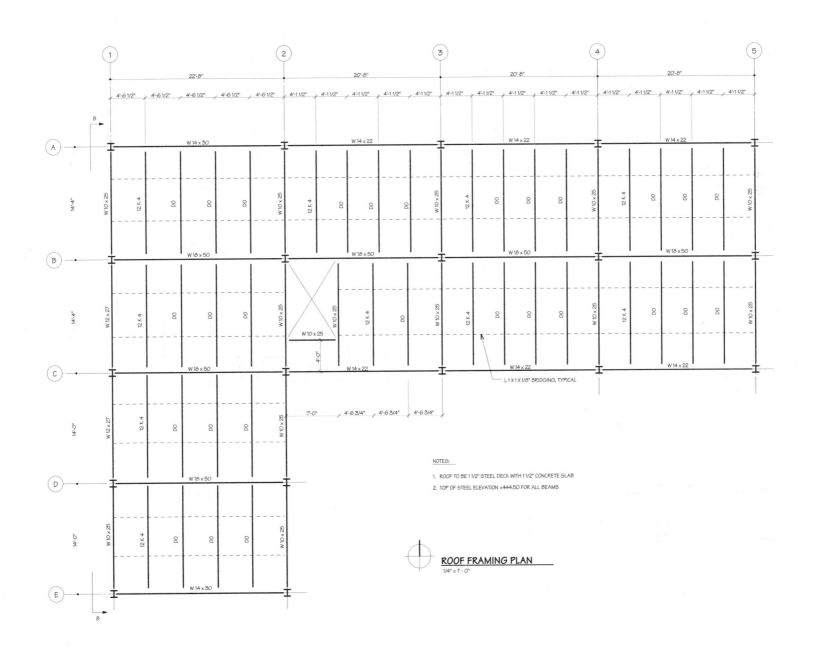

ROOF FRAMING PLAN
1/4" = 1' - 0"

NOTES:

1. ROOF TO BE 1 1/2" STEEL DECK WITH 1 1/2" CONCRETE SLAB
2. TOP OF STEEL ELEVATION +444.50 FOR ALL BEAMS

# Steel Column Schedules

Schedules are not used much for wood frame structures but are very valuable for steel frames. Many schedules are used to provide information about the structure but easily the most often used and necessary is the column schedule.

A column schedule is similar to the pier and footing schedules shown earlier in that it is a table of information about each column. It differs in that it is also a schematic or symbolic drawing of the relationship between the columns, their bases, and the floor levels. Column bases are called **base plates** and help level the column and distribute its load over a larger area.

There are minor variations in the way column schedules can be drawn. The one shown here encompasses the most common features. Similar columns are grouped together and identified across the top of the schedule. Their structural shape and size is listed below them. Note that the column marks are from the grid system.

A thick, vertical line representing the column itself is drawn below the grouping of column marks. This line crosses horizontal lines that represent floor and roof elevations. These elevations are noted above and to the left of each horizontal line.

A short horizontal line at the bottom of the vertical line represents the base plate. If it falls above or below the floor line it is drawn and noted so. The same is true of the top of the column. The final row of information at the bottom of the schedule is titled "base plate," which gives the size of this element.

The title for the schedule runs along the top as shown. If at all possible, this schedule should be located on the same sheet as the framing plan for easy reference.

| COLUMN SCHEDULE | | | | |
|---|---|---|---|---|
| COLUMN MARK | A1, A5, C5 E1, E2 | A2, A3, A4 B1, B5 C1, C3, C4 D1, D2 | C2 | B2, B3, B4 |
| FLOOR ELEVATION | W 10 x 22 | W 10 x 33 | W 10 x 39 | W 10 x 45 |
| ROOF  ELEV. 444.96' | | | | |
| | -5 1/2" | -5 1/2" | -5 1/2" | -5 1/2" |
| FIRST FLOOR  ELEV. 433.50' | -8" | -8" | -8" | -8" |
| BASE PLATE | 12" x 12" x 3/4" | 12" x 12" x 7/8" | 12" x 12" x 7/8" | 12" x 12" x 1" |

# STEEL SECTIONS/ELEVATIONS

Except for more complex steel projects, structural sections/elevations are not needed. The structure can usually be described with the framing plans, schedules, and details. A full structural section or elevation is a way, however, of showing how all the structural members fit together to form a complete structure.

Essentially all the conventions that apply to a steel framing plan also apply to full sections or elevations. The members are drawn as representational lines and noted, the grid is superimposed with tags, dimensions and/or elevations are given, and miscellaneous notes are added.

Partial sections are those that focus on one column and the beams that are connected to it. They are drawn at a larger scale and therefore allow the drafter the opportunity of drawing them as actual representations of the members involved. In reality they are not any different than structural details, which will be shown next.

As with full sections, each member must be identified with a note. Key elevations and dimensions must be provided, as well as any grid tags. If possible, any sections should be located on the framing plan from which it is constructed and cross referenced.

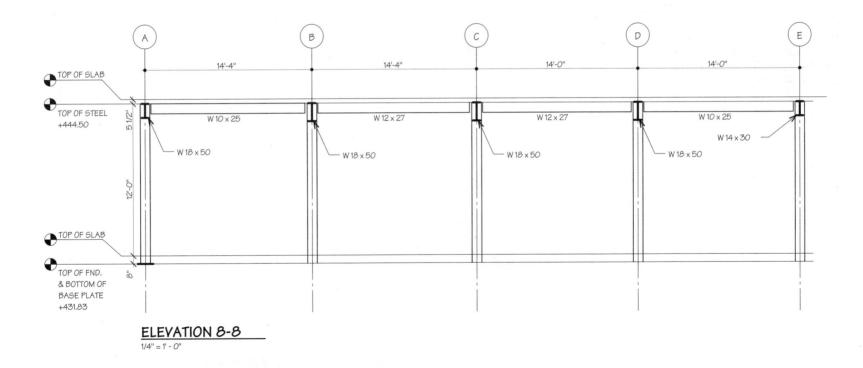

**ELEVATION 8-8**
1/4" = 1' - 0"

# STEEL DETAILS

The final drawings found in a structural set are details. Details take any form we have already seen — plan, section, or elevation — because they are simply enlargements of those drawings. They are developed exactly the same way as the parent drawing and follow the same conventions.

The key purpose of details is to show how the members are attached to each other. Secondarily they show the relationship between parts in terms of the exact position and location of one member to the next.

What needs to be included in a detail is covered in the following steps:

- draw reference lines for location (grid lines with column tags and floor or top of steel elevations)
- draw and label all members
- dimension members (unless they fall on center lines, which most do)
- show and note the method of connection

Structural steel members are connected by bolting or welding. Bolting is shown here because it is rather straightforward. Welding, on the other hand, involves a symbolic notation system. It is a topic more suited for advanced courses and texts.

The drafter needs to show and note where the connection occurs in the detail and what type of connection it is. This is illustrated in the example shown at right.

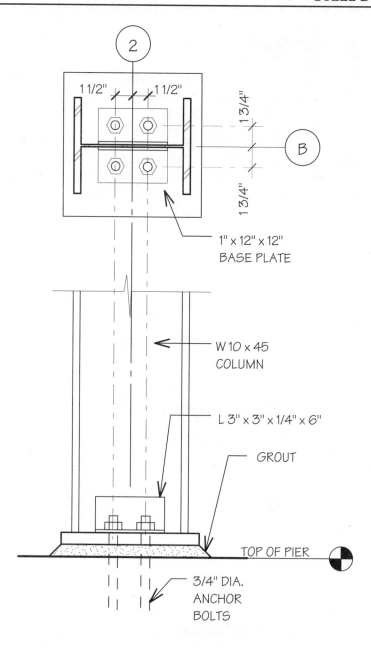

## BASE PLATE DETAIL
1 1/2" = 1' - 0"

# CONCRETE FRAMING

Concrete frames are much more complex than either wood or steel and cannot be covered in an introductory text such as this. A simple introduction is presented here to give an idea as to their complexity.

To begin with, there are eight major ways of constructing concrete floors: one-way flat slabs, two-way flat plates, two-way flat slabs with drops, one-way beam and slab, two-way beam and slab, skip joist, one-way joist slab, and two-way joist (waffle) slab.

Floors and columns can be combined into one drawing for steel and wood structures, but concrete structures are too complicated for that. The configuration for each is shown most clearly by drawing a plan for slabs and another for columns.

On top of that, there must be separate plans for concrete work and reinforcing work. Trying to show the shape of the concrete and the reinforcing within it would be too much information for one plan. Plans that show concrete work are called *engineering drawings*, and those that show reinforcing are called *placing drawings*.

This means four drawings are required for concrete framing to show what can be shown on one drawing for steel and wood systems:

- column engineering drawings
- column placing drawings
- floor engineering drawings
- floor placing drawings

Given eight different concrete structures and four different plans for each, there are thirty-two types of structural concrete plans a drafter must be familiar with.

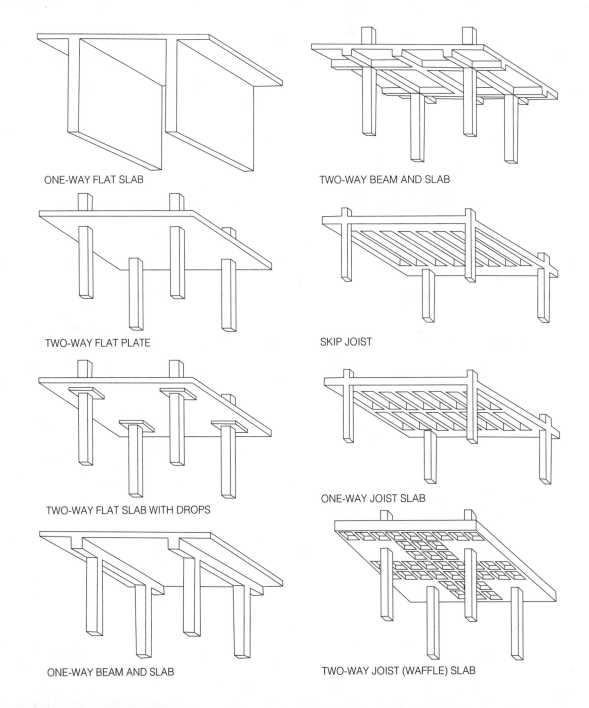

ONE-WAY FLAT SLAB

TWO-WAY BEAM AND SLAB

TWO-WAY FLAT PLATE

SKIP JOIST

TWO-WAY FLAT SLAB WITH DROPS

ONE-WAY JOIST SLAB

ONE-WAY BEAM AND SLAB

TWO-WAY JOIST (WAFFLE) SLAB

# KEY TERMS

sub-structure
foundation
super-structure

framing
pier
pilaster
footing
reinforcing
crawl space

frost line
mark
dowel
tie
joist
on center

header
glue laminated
cross bracing
cross bridging
sub-floor
plywood

underlayment
bay
base plate
engineering drawings
placing drawings

# PRACTICE EXERCISE

For the buildings from the practice site plan of the last chapter, complete the following drawings:

- foundation plans
- floor framing plans
- roof framing plans
- steel column schedule
- partial steel section
- 3 details

The wood frame building is 20' x 20'. The foundation is 12" concrete block on 2' wide by 1' deep concrete footings with 2 continuous #5 reinforcing bars. The foundation wall is 6 courses deep forming a crawl space with 4" slab.

The floor is made with 2 x 8s @ 16" o.c. that are supported mid-way with a W8 x 20 steel beam. A 3" pipe column supports the beam at mid span and is supported with a 3' x 3' x 1' deep concrete footing.

Provide additional joists where needed and an opening for a mechanical shaft.

The one story masonry building is 60' x 80'. Its foundation is poured concrete. The sketch at right supplies the necessary information to complete the drawings.

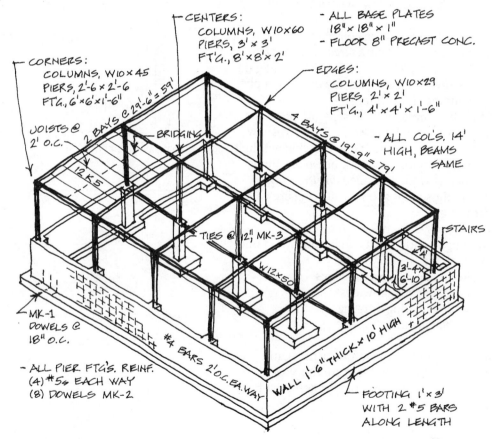

- PERIMETER BEAMS, W12 x 50
- INTERIOR BEAMS, W10 x 112

- ROOF, 2" STL. DECK WITH 2" CONC. TOPPING

CENTERS:
COLUMNS, W10 x 60
PIERS, 3' x 3'
FT'G., 8' x 8' x 2'

- ALL BASE PLATES 18" x 18" x 1"
- FLOOR 8" PRECAST CONC.

CORNERS:
COLUMNS, W10 x 45
PIERS, 2'-6 x 2'-6
FT'G., 6' x 6' x 1'-6"

EDGES:
COLUMNS, W10 x 29
PIERS, 2' x 2'
FT'G., 4' x 4' x 1'-6"

2 BAYS @ 29'-6" = 59'

4 BAYS @ 19'-9" = 79'

- ALL COL'S. 14' HIGH, BEAMS SAME

JOISTS @ 2' O.C.

12 K 5

BRIDGING

TIES @ 12", MK-3

STAIRS

W12 x 50

MK-1
DOWELS @ 18" O.C.

#4 BARS 2' O.C. EA. WAY

WALL 1'-6" THICK x 10' HIGH

- ALL PIER FTG's. REINF.
(4) #5 EACH WAY
(8) DOWELS MK-2

FOOTING 1' x 3' WITH 2 #5 BARS ALONG LENGTH

# APPENDIX

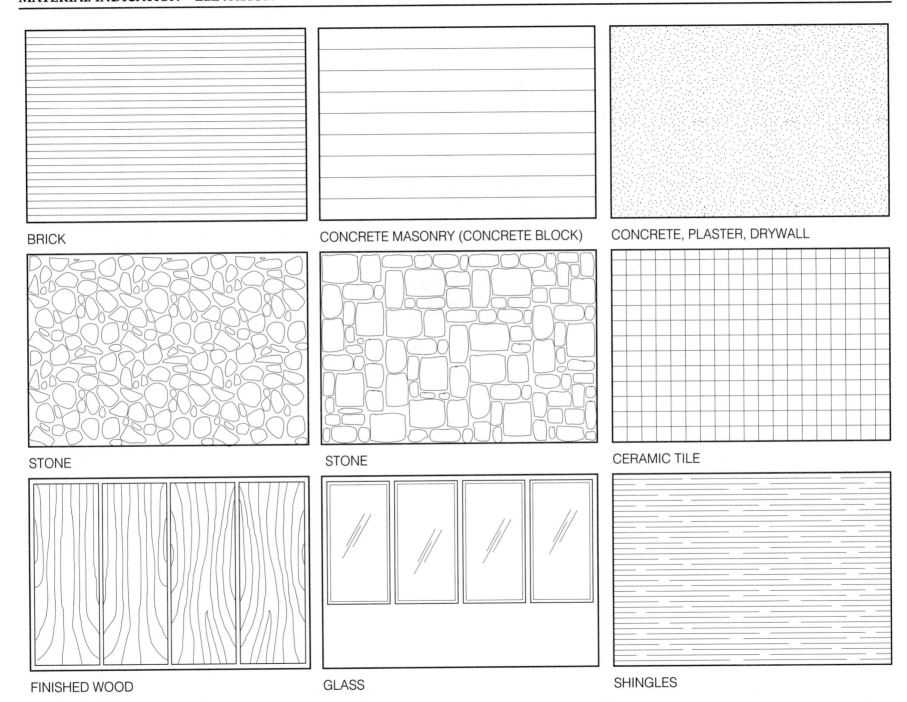

BRICK

CONCRETE MASONRY (CONCRETE BLOCK)

CONCRETE, PLASTER, DRYWALL

STONE

STONE

CERAMIC TILE

FINISHED WOOD

GLASS

SHINGLES

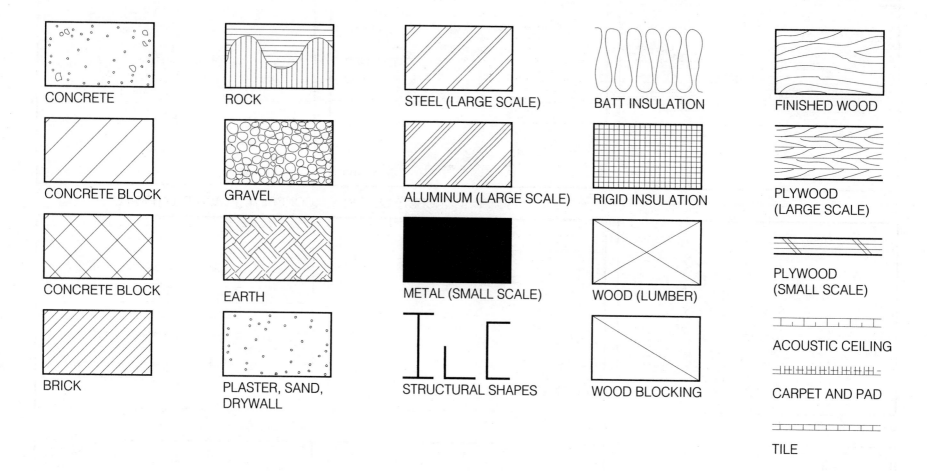

CONCRETE

ROCK

STEEL (LARGE SCALE)

BATT INSULATION

FINISHED WOOD

CONCRETE BLOCK

GRAVEL

ALUMINUM (LARGE SCALE)

RIGID INSULATION

PLYWOOD
(LARGE SCALE)

CONCRETE BLOCK

EARTH

METAL (SMALL SCALE)

WOOD (LUMBER)

PLYWOOD
(SMALL SCALE)

BRICK

PLASTER, SAND,
DRYWALL

STRUCTURAL SHAPES

WOOD BLOCKING

ACOUSTIC CEILING

CARPET AND PAD

TILE

| | | | | | | | |
|---|---|---|---|---|---|---|---|
| ABV | above | CAB | cabinet | DIAG | diagonal | FH | fire hydrant |
| AP | access panel | CAL | caliper | DIA | diameter | FPL | fireplace |
| AC | acoustic | CPT | carpet | DIM | dimension | FP | fireproof |
| ACT | acoustic tile | CSMT | casement | DR | door | FLG | flashing |
| AC | acre(s) | CI | cast iron | DH | double hung | FLX | flexible |
| ADD | addendum | CIPC | cast-in-place concrete | DS | downspout | FLR | floor(ing) |
| ADH | adhesive | CB | catch basin | D | drain | FD | floor drain |
| ADJ | adjacent | CK | caulk, caulking | DT | drain tile | FTG | footing |
| AGG | aggregate | CLG | ceiling | DWG | drawing | FND | foundation |
| A/C | air conditioning | CHT | ceiling height | DF | drinking fountain | FR | fram(ed)(ing) |
| ALT | alternate | CEM | cement | DMH | drop manhole | FUR | furring(ed) |
| AL | aluminum | CM | centimeter(s) | DW | dumbwaiter | | |
| ANC | anchor | CER | ceramic | | | G | gas |
| AB | anchor bolt | CMT | ceramic mosaic tile | EA | each | GALV | galvanized |
| APX | approximate | CT | ceramic tile | EF | each face | GA | gauge |
| ARCH | architect, architectural | CLF | chain link fence | EW | each way | GC | general contractor |
| AD | area drain | CIR | circle, circular | E | east | GLB | glass block |
| ASPH | asphalt | CIRC | circumference | ELEC | electric, electrical | GCMU | glazed concrete masonry unit |
| AT | asphalt tile | CO | cleanout | EP | electric panelboard | | |
| AUTO | automatic | CLR | clear, clearance | EL | elevation | GST | glazed structural tile |
| | | COL | column | ELEV | elevator | GB | grab bar |
| B&B | balled and burlapped | CT | column tie | EQ | equal | GD | grade |
| BSMT | basement | COMB | combination | EQP | equipment | GVL | gravel |
| BRG | bearing | CONC | concrete | EXCA | excavate | GT | grout |
| BPL | bearing plate | CMU | concrete masonry unit | EXG | existing | GYP | gypsum |
| BEL | below | CONST | construction | EB | expansion bolt | | |
| BM | benchmark | CONT | continuous | EAP | exposed aggregate paving | HBD | hardboard |
| BT | bent | CONTR | contractor | EXT | exterior | HDW | hardware |
| BET | between | CLL | contract limit line | | | HWD | hardwood |
| BIT | bituminous | CJT | control joint | FB | face brick | HDR | header |
| BLK | block | CMP | corrugated metal pipe | FOC | face of concrete | HTG | heating |
| BLKG | blocking | CTR | counter | FOF | face of finish | HVAC | heating/ventilating/air conditioning |
| BD | board | CFL | counterflashing | FOM | face of masonry | | |
| BS | both sides | CS | countersunk | FOS | face of studs | HT | height |
| BW | both ways | CRS | course(s) | FF | factory finish | HWY | highway |
| BOT | bottom | CFT | cubic foot (feet) | FN | fence | HC | hollow core |
| BC | bottom of curb | CYD | cubic yard(s) | FBD | fiberboard | HM | hollow metal |
| BS | bottom of slope | | | FGL | fiberglas | HOR | horizontal |
| BRK | brick | DP | damproofing | FIN | finish(ed) | HWH | hot water heater |
| BLDG | building | DEM | demolish, demolition | FF(E) | finished floor (elevation) | | |
| BUR | built-up roofing | DTL | detail | FG | finished grade | | |

| | | | | | | | |
|---|---|---|---|---|---|---|---|
| IN | inch | MC | medicine cabinet | PERI | perimeter | SG | sheet glass |
| INCL | include(d)(ing) | MED | medium | PLAS | plaster | SH | shelf, shelving |
| INL | inlet | MBR | member | PLAM | plastic laminate | SIM | similar |
| ID | inside diameter | MET | metal | PL | plate | SKL | skylight |
| INS | insulate, insulation | M | meter | PWD | plywood | SC | solid core |
| INT | interior | MM | millimeter | PT | point | S | south |
| INV | invert | MWK | millwork | PCC | point of compound | SPL | special |
| IPS | iron pipe size | MIN | minimum | | curvature | SPEC | specification(s) |
| | | MIR | mirror | PC | point of curvature | SQ | square |
| JC | janitor's closet | MISC | miscellaneous | PVC | polyvinyl chloride | SQB | square beam |
| JT | joint | MON | monument | PCP | porous concrete pipe | SQCOL | square column |
| JF | joint filler | MULL | mullion | PCC | precast concrete | SST | stainless steel |
| J | joist | | | PFN | prefinished | STD | standard |
| | | NAT | natural | PL | property line | STA | station |
| KPL | kick plate | NF | near face | | | STL | steel |
| KIT | kitchen | NOM | nominal (dimension) | QT | quarry tile | STO | storage |
| KO | knockout | N | north | | | SD | storm drain |
| | | NIC | not in contract | RAD | radius | SI | storm inlet |
| LBL | label | NTS | not to scale | RB | rectangular beam | ST | storm sewer |
| LAB | laboratory | NO | number | RC | recangular column | STR | structural |
| LAD | ladder | | | REF | reference | SUS | suspended |
| LB | lag bolt | OC | on center | REFR | refrigerator | SYS | system |
| LAM | laminate | OPG | opening | RE | reinforce, reinforcing | | |
| LAV | lavatory | OJ | open-web joist | RCP | reinforced concrete pipe | TEL | telephone |
| L | length | OPP | opposite | REQ'D | required | TV | television |
| LT | light | OPH | opposite hand | REV | revision | TC | terra cotta |
| LWC | lightweight concrete | OD | outside diameter | ROW | right of way | TZ | terrazzo |
| LMS | limestone | OF | outside face | R | riser | THK | thick(ness) |
| LTL | lintel | QA | overall | RD | roof drain | THR | threshold |
| LVR | louver | OH | overhead | RFH | roof hatch | TPD | toilet paper dispenser |
| LP | low point | | | RFG | roofing | T&G | tongue and groove |
| | | PNT | paint(ed) | RM | room | TC | top of curb |
| MB | machine bolt | PNL | panel | RO | rough opening | TSL | top of slab |
| MH | manhole | PTD | paper towel dispenser | | | TS | top of slope |
| MFR | manufacturer | PTR | paper towel receptor | SAN | sanitary sewer | TST | top of steel |
| MRB | marble | PAR | parallel | SCH | schedule | TW | top of wall |
| MAS | masonry | PK | parking | SNT | sealant | TB | towel bar |
| MO | masonry opening | PBD | particleboard | SEC | section | T | tread |
| MTL | material | PTN | partition | SSK | service sink | TYP | typical |
| MAX | maximum | PV | pave, paving | SHTH | sheathing | | |
| MECH | mechanical | PVMT | pavement | SHT | sheet | | |

| | |
|---|---|
| UC | undercut |
| UD | underdrain |
| UNF | unfinished |
| UR | urinal |
| USGS | US geologic survey |
| | |
| VB | vapor barrier |
| VAR | varies |
| VNR | veneer |
| VERT | vertical |
| VC | vertical curve |
| VIN | vinyl |
| VB | vinyl base |
| VF | vinyl fabric |
| VT | vinyl tile |
| | |
| WSCT | wainscot |
| WTW | wall to wall |
| W | water |
| WC | water closet |
| WP | waterproofing |
| WWF | welded wire fabric |
| W | west |
| W | wide, width |
| WIN | window |
| WG | wire glass |
| WM | wire mesh |
| W | with |
| WO | without |
| WD | wood |
| WB | wood base |
| WI | wrought iron |
| | |
| YD | yard drain |
| | |
| @ | at |
| ' | feet |
| " | inches |
| # | number |
| # | pound |
| Ø | round diameter |

| CROSS REFERENCING, SYMBOL TYPE | SIZE | EXAMPLE |
|---|---|---|
| **DIMENSIONS**<br><br>• MASONRY TO FACES<br>• ALL OTHERS TO CENTER LINE<br><br>• DOTS FOR COLUMN CENTER LINES<br>• ARROWS FOR NARROW DIMENSIONS<br>• SLASHES FOR ALL OTHERS | | MASONRY · ALL OTHERS · COLUMN CENTER LINE |
| **BUILDING CROSS SECTION**<br><br>• LETTERS: A-A, B-B, ETC.<br>• ON PLANS ONLY | 3/8" DIA. | A / A-6 · LETTER · SHEET WHERE FOUND · B / A-7 · B / A-7 · A / A-6 |
| **DETAIL SECTION**<br><br>• NUMBERS: 1, 2, 3, ETC.<br>• ON PLANS AND ELEVATIONS | 3/8" DIA. | 1 / A-8 · NUMBER · SHEET WHERE FOUND · 2 / A-8 · TAIL ARROW |
| **DETAIL ENLARGEMENT**<br><br>• NUMBERS 1, 2, 3, ETC.<br>• ON PLANS, ELEVATIONS, AND SECTIONS | AS REQUIRED<br><br>3/8" DIA. | 4 / A-8 · NUMBER · SHEET WHERE FOUND |

| CROSS REFERENCING, SYMBOL TYPE | SIZE | EXAMPLE |
|---|---|---|
| COLUMN LINE | 3/8" DIA. | (C) ———— – – – – – – |
| DETAIL TITLE<br><br>• NUMBERS: 1, 2, 3, ETC. | 1/2" DIA. | **BASE DETAIL**<br>1 1/2" = 1' - 0"<br><br>NUMBER<br>(5)<br>(A-3)<br>LOCATION OF PARENT DRAWING |
| INTERIOR ELEVATIONS<br><br>• ON PLANS ONLY<br>• LETTERS AND NUMBERS, E.G. ELEVATION G1, G2, G3 | 3/8" DIA. | 2<br>1 (G / A-9) 3 ← INDICATES ELEVATION G2<br>← SHEET WHERE DRAWING MAY BE FOUND<br>← NO ELEVATION, NO ARROW |
| ROOM NUMBER AND TITLE<br><br>• ON PLANS ONLY | 1/4" x REQUIREMENT | **OFFICE**<br>328 |
| DOOR NUMBER<br><br>• ON PLANS, ONLY | 3/8" DIA. | (213) |
| ELEVATION MARK<br><br>• ON ELEVATIONS<br>• ON SECTIONS AND DETAILS AS REQUIRED | 1/4" DIA. | TOP OF STEEL<br>+ 25' - 4 1/2" |

| CROSS REFERENCING, SYMBOL TYPE | SIZE | EXAMPLE |
|---|---|---|
| **NORTH ARROW**<br><br>• ON PLANS AND SITE PLANS ONLY<br>• ADD TRUE NORTH LINE WHEN ORIENTATION IS SKEWED | 3/4" TO<br>1" DIA. | 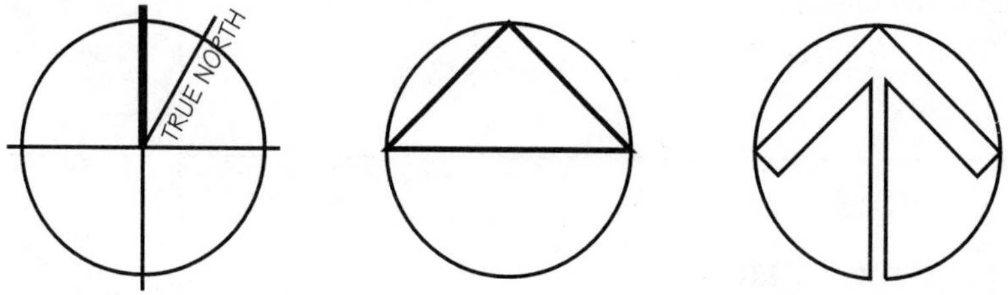 |
| **LETTERING**<br><br>• GENERAL NOTES<br>   ARROWS TO FACE<br>   DOTS TO AREAS | 3/32" | |
| • DRAWING SUBTITLES OR ROOM NAMES, ETC. | 3/16" | |
| • DIMENSIONS OR SPECIAL NOTES | 1/8" | |
| • DRAWING TITLE | 1/4" | |
| • SCALE | 1/8" | |
| • SHEET NUMBER, MACHINE MADE OR HAND OUTLINE | 1/2" TO 1" | |

CONCRETE LINTEL

HEAD

INSULATING GLASS

ALUMINUM FRAME

CONCRETE BLOCK

SILL

4' - 8"

# WINDOW DETAILS
3" = 1' - 0"

A-1

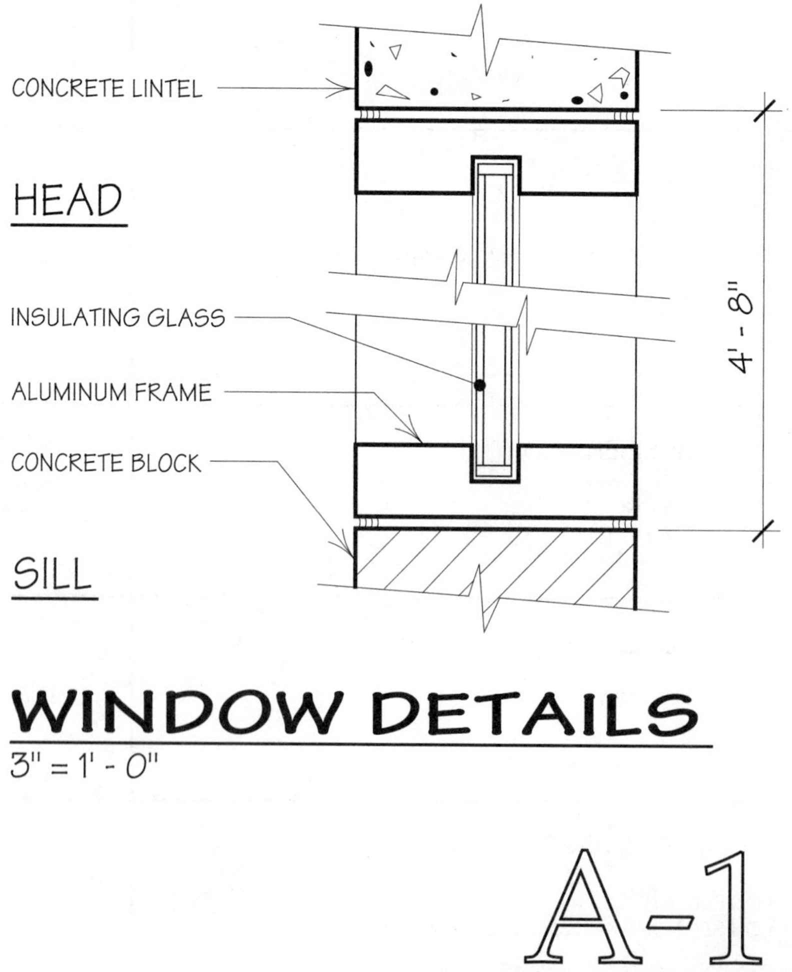

# INDEX

# Colophon

This text was written, edited, and composed on a desktop publishing system using Apple Macintosh computers.

The pages were composed in Aldus PageMaker®. General illustrations were drawn with Adobe Illustrator®.

CAD drawings were produced with ArchiCAD®, saved as plot files, and imported into Plotmaker®. Once in Plotmaker, the files were saved as Encapsulated PostScript® (EPS) illustrations. Both ArchiCAD and PlotMaker are products of Graphisoft. PostScript was developed by Adobe Systems Incorporated.

Text type is Hiroshige. Occasional other type is Helvetica, Palatino, and Zapf Dingbats. Type for general illustrations is Helvetica Light. Simulated hand lettering for drawings is Tekton.

Hand drawn illustrations were scanned and converted to electronic files. Output was directly to negative for printing.